T0224976

Traces of the Ice Age

Wolfgang Fraedrich

Traces of the Ice Age

Landscape Forms in Central Europe

 Springer

Wolfgang Fraedrich
Hamburg
Hamburg, Hamburg, Germany

ISBN 978-3-662-65888-8   ISBN 978-3-662-65886-4   (eBook)
https://doi.org/10.1007/978-3-662-65886-4

This book is a translation of the original German edition „Spuren der Eiszeit" by Fraedrich, Wolfgang, published by Springer-Verlag GmbH, DE in 2016. The translation was done with the help of artificial intelligence (machine translation by the service DeepL.com). A subsequent human revision was done primarily in terms of content, so that the book will read stylistically differently from a conventional translation. Springer Nature works continuously to further the development of tools for the production of books and on the related technologies to support the authors.

This Springer imprint is published by the registered company Springer-Verlag GmbH, DE, part of Springer Nature.
The registered company address is: Heidelberger Platz 3, 14197 Berlin, Germany

Preface

Around 10% of the Earth's land surface – just under 15 million km^2 – is currently covered by glacial ice. But within the last 2.6 million years, there have been periods in which the ice cover was far greater. Large parts of northern and southern Germany have also been repeatedly covered by a thick sheet of ice, have been shaped by meltwaters or by winds in a predominantly hostile climate. This book is intended to sharpen the view of the landscapes that were formed by the ice and from the ice. The search for traces begins on our doorstep, leads to the present glacial landscapes of the earth and back to the cold periods of the earth's history. Even though the title of the book uses the more familiar term "ice ages", the scientifically more precise term "cold ages" will be used in the following chapters.

The book is an updated new edition of the work first published in 1996. Especially in the field of research on the climate of prehistory, science today is a significant step further than 20 years ago, which is adequately reflected by this new edition.

The view into the future is interesting because in the age of globalization, economic man is increasingly becoming a climate factor. The question of his contribution to climate development is still assessed differently. It is therefore necessary to provide facts and indications for an assessment of current developments.

The earth's climate has always experienced fluctuations, but there have not always been ice ages like the one we have been living in for about 2.6 million years and which will continue for several million years. Why this is so will also be clarified in this book.

Hamburg, Germany   Wolfgang Fraedrich
October 2015

Contents

Searching for Clues on Our Doorstep

1

Wolfgang Fraedrich

Abstract

It is hardly imaginable today, but there were periods of several 10,000 years in the recent history of the earth during which large parts of Europe were covered by mighty ice caps.

In this chapter, a basic understanding of cold-formed landscapes is first to be laid by the dimensions of cold-age glaciation are made clear. It is explained that glaciers are capable of transporting unimaginably large quantities of rock. It illustrates, which regions, e.g. in Germany, have been shaped in a special way by cold periods – i.e. by glacial ice, but also by meltwater and periglacial processes. And finally, it raises awareness of the fact that cold periods are an expression of climate fluctuations and changes that have always been typical of the Earth, and that humans have a legitimate interest in gaining an outlook on the climate in the near future.

It is hard to imagine today, but there were periods of several 10,000 years in the recent history of the earth during which large parts of Europe were covered by mighty ice caps.

The Nordic **ice sheet**, which originated on the Fennoscandian peninsula and flowed from there over several 100 km mainly towards the south, reached its greatest extent during the older Saale ice age. During the last **cold period** alone, which ended about 12,000–10,000 years before today, the Nordic inland ice reached a maximum thickness of about 2500 m. This was an enormous load on the northern European mainland.

W. Fraedrich (✉)
Hamburg, Hamburg, Hamburg, Germany
e-mail: wolfgang.fraedrich@hamburg.de

Fig. 1.1 Moraine landscape in the northeast of Hamburg

On the way from Scandinavia to Central Europe, the repeatedly advancing ice masses, some of which had melted back almost completely in longer warm periods of more than 20,000 years, moved billions of tons of rock. The Scandinavian mountains were literally "planed" in the process, and the **eroded debris** was transported, among other places, as far as north-central Europe on its way through the Baltic Sea basin, which was still being scraped out by the force of the ice. The entire North German Lowland (Fig. 1.1), the entire **shelf of** the North Sea, large parts of the British Isles, the north of the Netherlands, but also large parts of Poland, the Baltic and northwestern Russia were covered by the ice **sheet**. And everywhere where the post-cold-age seas did not advance, the earth's surface is essentially built up of sedimentary material from the Nordic ice sheet.

With regard to the forces and processes as well as the temporal sequence, comparable but in their spatial dimensions far less significant ice advances occurred in the Alpine region and Alpine foothills. In the Alps – mainly due to their altitude – a self-glaciation could develop; here also mighty glaciers built up. They carried away material from the inner Alpine (high) mountain region and deposited it mainly in the foreland areas.

The relief-forming forces of the flowing ice and meltwater and the fluctuating climatic conditions, which in turn shaped the ice margin as a dynamic system, resulted in a highly diverse landscape. Above all, the many lakes, e.g. in Holstein

Switzerland, Lauenburg or the Mecklenburg Lake District, but also in the German Alpine foothills, document the young "glacial history" of Germany. The diversity of the Baltic Sea coastal region is also primarily a result of the effectiveness of the flowing inland ice. Not least because of their varied landscapes, the regions shaped by the ice are also preferred tourist destinations.

If one imagines the extreme north and the extreme south of Germany covered with ice, one almost "automatically" comes to the question of what the landscapes in between might have looked like. Well, it will have been a very inhospitable region – measured by today's standards. Short, rather cool summers and long winters marked the course of the year. There were no extensive forests, it was too cool for that, and the **vegetation period** was too short. In the tundra zone, even if the level of human development had made it possible at that time, it would never have been possible to grow crops profitably. At the heights of the low mountain ranges there was even an **icy climate**.

But even in this landscape there were landscape changes, triggered by natural forces and processes controlled by atmospheric conditions.

If we want to learn something about how glacial ice and its meltwater shape surface forms, and if we want to get clues about how landscapes change at the edge of the ice, it is not necessary to travel far, we can find the traces on our doorstep. Comparison with corresponding forms in other landscapes shaped by ice helps us to understand and classify the relationships today.

Quite independently of the diversity of forms, the phenomenon of the "cold period" is also coming to the fore again for another reason: scientific research in the Earth's glacial regions, as we now know, provides very important information about the **climate history of** the recent **modern era**. It is hoped that this kind of "search for traces" will provide clues for the most accurate possible prognosis for the climate of the future, especially that of the next 100–200 years.

However, while the glacial landscapes as a whole are now quite well researched, climate research is still in its infancy. This will also be a major focus of scientific research in the coming decades.

Around the Glacier

2

Wolfgang Fraedrich

Abstract

The landscapes formed by the ice have their charm not only because of their variety of forms, they are simply beautiful and varied landscapes, no matter where we are in Central Europe. And with just a little trained eye, the history of the formation of such ice-formed landscapes can be understood even by the interested layman. In order to have the required basis for this, it is necessary to take a closer look at the phenomenon of glaciers. Only when the connections to be worked out are clear, will one learn to understand landscapes. But in order to grasp and understand the phenomenon of glaciers, one must look into the present, namely into those regions where there is perpetual ice. In this context the information "Around the Glacier" is an important basis for the understanding of glacial and fluvioglacial forces and processes. Here you learn how glaciers are formed and what conditions must exist for ice formation, you learn that glaciers have different regional names and that glaciers are typified according to extent and structure and that each glacier has different characteristics. There are informations about the conditions of the glacier's ice flows. You learn that glaciers vary in extent and ice mass depending on the prevailing climate and its fluctuations. Finally you will get information about the fact that the registration of current and the reconstruction of historical glacier fluctuations are essential components of glaciological research.

W. Fraedrich (✉)
Hamburg, Hamburg, Hamburg, Germany
e-mail: wolfgang.fraedrich@hamburg.de

2.1 Learning to Understand Glaciers

Glaciated areas no longer exist in Central Europe, with the exception of the glaciers at high altitudes in the Alps, but you can still track the ice here. The landscapes formed by the ice have their charm not only because of their variety of forms, they are simply beautiful and varied landscapes, no matter where we are in Central Europe. And with just a little trained eye, the history of how such ice-formed landscapes came into being is also comprehensible to the interested layman.

In order to have the necessary basis for this, it is first necessary to deal more closely with the phenomenon of glaciers. Only when the interrelationships to be worked out in the process are clear will we learn to understand landscapes. But in order to grasp and understand the phenomenon of glaciers, we must of course also look into the present, namely into those regions where there is perpetual ice.

2.2 How Glaciers Form

To understand glacial processes in the broadest sense, knowledge of the development, structure and shaping forces of glacial ice is therefore required.

But the consideration should already begin with the snow, because the snow is the starting material for the glaciers. Snowfall is registered on about 80% of the continental surfaces. Where the snow remains over a longer period of time, it is withdrawn from the earthly **water cycle**. If it then even forms ice, a huge freshwater reservoir is created, the largest on our planet (Table 2.1).

Along with water, ice and wind, snow is one of the forces that shape the earth's surface. Its clearest traces are left by avalanches in the mountains; they are a typical process of **nivation**.

▶ **Nivation** Nivation is the term scientists use to describe all processes in which the relief is shaped by the direct action of the snow on the ground as well as by movement, pressure and (snow) meltwater.

Table 2.1 Water quantities of the earth – water reserves

	in km^3	in %
Total water volume	1,385,984,610	100.0
Water in the oceans	1,338,000,000	96.5
Groundwater[a]	23,400,000	1.7
Water in glacial ice and snow	24,064,000	1.74
Water in salt and freshwater lakes	176,400	0.013
Soil moisture	16,500	0.001
Water in the atmosphere	12,900	0.001
Stream water	2120	0.0002
Water in the permafrost ground ice	300,000	0.022
Water in marshes	11,470	0.0008
Water stored by living organisms	1120	0.0001

[a]Without Antarctic groundwater reserves of about 2,000,000 km^3

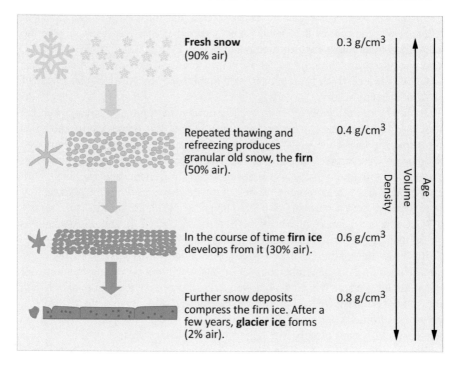

Fresh snow (90% air) — 0.3 g/cm³

Repeated thawing and refreezing produces granular old snow, the **firn** (50% air). — 0.4 g/cm³

In the course of time **firn ice** develops from it (30% air). — 0.6 g/cm³

Further snow deposits compress the firn ice. After a few years, **glacier ice** forms (2% air). — 0.8 g/cm³

Density Volume Age

Fig. 2.1 From snow to ice. (© Wolfgang Fraedrich)

Snowfall and the formation of a snow cover are determined by two factors, temperature and the amount of snow. Both factors are primarily dependent on the climate and the vertical **altitude gradation** in the mountains, so that essentially four areas can be distinguished:

- The region above the climatic **snow line**, where – regardless of latitude – more snow falls than precipitation than is lost through evaporation and melting processes.
- Areas with permanent snow cover beyond the −3 °C isotherm.
- Areas with an irregular snow cover, essentially limited to the winter and spring months. This zone extends between the −3 °C and + 4 °C isotherms.
- The area up to the equatorial snow line for which snowfall is recorded only extremely sporadically and in which a permanent snow cover does not form.

Due to this division and the present water-land distribution of the earth, the fact arises that the snow-covered regions of the earth extend essentially on the northern hemisphere of the earth.

Glacial ice forms from snow (Fig. 2.1). The ratio of snow to ice is 80:1, i.e. 80 cm of new snow are required to form only 1 cm of glacial ice, or: 240,000 m (= 40 km) of new snow were required to form the approximately 3000 m thick ice cover in central

Greenland. Of course, this amount of snow does not fall in a few, but rather in thousands of years. Since the annual snow accumulation is low, especially in the polar regions with relatively little precipitation, very long periods of time pass before thicker ice sheets are formed. A borehole in the Greenland ice sheet (cf. Chap. 9) has shown that the ice, which is more than 3000 m thick, covers about 250,000 years of the Earth's recent history.

The formation of ice occurs most rapidly during the summer months, when the snow thaws and refreezes (usually overnight). During this process, also known as **regelation**, the fine-jetted snow crystals transform into the granular **firn**. Renewed precipitation exerts pressure and the firn is compressed. Ingressing meltwater enlarges the firn grains, which also reduces the air content, and eventually firn becomes glacial ice. According to the definition, this happens when the permeability for air becomes zero at a density of 0.82–0.85 g/cm^3. While fresh snow still contains 90% air, the air content of bluish glacier ice is only 2%, and the density is correspondingly different (snow: 0.1 g/cm^3, glacier ice: 0.9 g/cm^3).

However, the conditions under which glacier ice forms vary from region to region, depending on the temperature of the snow and the prevailing humidity. At higher latitudes, where it is colder and humidity is lower, ice formation takes a long time. Studies have shown that in northwest Greenland, for example, the transformation process from snow to glacial ice takes about 100 years. For parts of Alaska, on the other hand, values of only 3–5 years have been determined.

This ice (Figs. 2.2 and 2.3) appears light to dark blue, a colour effect that is enhanced in intense sunlight. The reason for this is that ice is physically constructed in such a way that it reflects mainly short-wave, blue light and swallows up the other colours.

The transformation (**metamorphosis**) of snow to glacial ice thus takes place in three phases:

- Compaction of the snow cover,
- Origin of firn (transformation of snow to firn),
- Glacial ice formation (transformation of firn to glacial ice).

2.3   Glacier, Ghiaccia, Jökull

The term "glacier" has been derived from the Latin word "glacies" – this means "ice". In the twelfth century, the Lower Glacier near Grindelwald was called "glacies inferior", which means "lower ice".

The German word "**Gletscher**" (**glacier**) **was** first mentioned in the Swiss chronicle by Petermann Etterlin in 1507. In the French-speaking world the word "glacier" has become established, in the Anglo-Saxon world the word "**glacier**". In Tyrol, the term "**Ferner**" (derived from the Old High German word firn = old) is common; the term "Firn", which is common in the Glarus Alps, is synonymous. Further east and south in the Alps, the term "**Kees**" (also derived from Old High German: ches = ice, cold) is common. In the Rhaeto-Romanic language area the

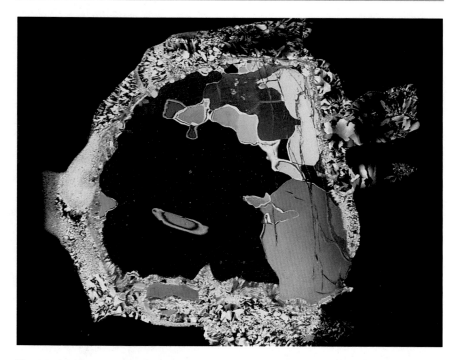

Fig. 2.2 Highly condensed ice crystals of Svinafellsjökull (Iceland) viewed in thin section under the microscope (image edge length 10 cm). (© Wolfgang Fraedrich)

glaciers are called "**Vedretta**", in the Italian Alps "Ghiaccia", in Graubünden "Wader" and in Valais "**Biengo**".

Outside the Alps, names like "**Bre**" ("Breen") or "**Brä**" (Norway), "**Jökull**" (Iceland) or "**Sermek**" (Greenland) are common, in the Russian language area one speaks of "ледник" (**Lednik** written in Latin transcription).

2.4 Glacier Types

Not every glacier looks the same. On the continent of Antarctica or on Greenland, which are the regions with the largest continuous ice cover and the largest ice thicknesses on Earth (cf. Chap. 3), there is **inland ice**; this is also referred to as the Greenland type. The largest glacier in Europe, Vatnajökull in Iceland, is called an **ice sheet**. Ice sheets are powerful, inland-ice-like covering glaciers, but they are far smaller than inland glaciers and also nowhere near as powerful. The Jostedalsbreen in Norway, the largest glacier on the European mainland, is a glacier of the Scandinavian type, a plateau glacier. In the high mountains of the world, all of which have glaciers above the climatic snow line, the **valley glacier** is predominant, and is also known as the alpine type.

Fig. 2.3 Highly condensed blue glacial ice at the base of Fox Glacier on the South Island of New Zealand. (© Sandra Limburg)

2.5   Structure and Properties of a Glacier

The structure of a glacier can be most clearly illustrated by the size of a manageable alpine glacier type. A view of the glacier tongue of the Pasterze (Fig. 2.4) below the Großglockner in the Hohe Tauern and a schematised longitudinal section (Fig. 2.5) are intended to illustrate the main features of the structure. Glaciers are divided into two main sections: Above the climatic snow line lies the **nutrient zone**, where the amount of new snow is higher than the amount of snow lost through evaporation and melting. The **mass balance** there is therefore positive. In the lower section, the snow melting **area, the mass balance is** negative. As a rule, the snowpack area is free of snow during the short summer.

Depending on the flow behaviour of the ice, the thickness of the ice and the relief of the bedrock, a glacier either has a closed surface or numerous **crevasses**.

Fig. 2.4 Pasterze – Grossglockner massif, Austria. (© Wolfgang Fraedrich)

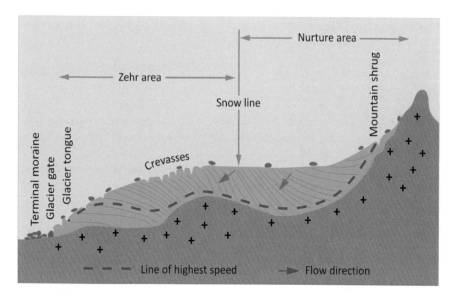

Fig. 2.5 Schematic longitudinal section through a valley glacier. (© Wolfgang Fraedrich)

Crevasses form especially when flowing over a terrain edge, but also at the edge of a glacier.

Due to the flowing movement of the ice and also due to melting processes, melt water is released, which – following the gradient in and under the ice – flows **subglacially** (= under the ice surface) towards the ice edge and emerges there at the **glacier gate**. While valley glaciers usually have only one glacier gate, the other glacier types have several. In many cases, plateau glaciers, ice sheets and inland ice bodies, which sometimes also cover higher mountain ranges, form glacier tongues towards the edge, for example Vatnajökull in southern Iceland, as documented by satellite images available via Google Earth®.

2.6   Glacial Ice Flows

Glacial ice flows, it is plastic, because the ice grains are built up in layers and they shift under pressure. Due to the change between liquid and solid aggregate state – this process is called **regelation** – the volume of the ice grains changes (they become larger) and the ice starts to move, following gravity.

▶ **Regelation** Regelation is the melting of glacier ice when pressure increases (freezing point decrease) and refreezing when pressure decreases (freezing point increase). Their cause is thus the change in melting point due to higher or lower pressure (by 7.3 °C per t/cm^2). An increase in pressure already occurs due to the surcharge and with compaction, and it is then amplified in the form of dynamic pressure, triggered either by movement in the glacier or also by obstacles at the base of the glacier.

For the layman it is difficult to imagine that glacier ice, unlike water, also flows upwards. This happens when the thrust of the ice – known as **firn field pressure** – in the nutrient area of the glacier is great enough to push the feeding area, e.g. a **glacier tongue**, uphill. The Sognefjord on the west coast of Norway, for example, was carved out by ice during the cold periods. At its deepest point, the ice was 1308 m below today's sea level. However, where the ice flowed into the area of today's North Sea at that time, a sea depth of only 175 m can be measured at present. An enormous difference in height of clearly more than 1000 m, which the ice has overcome!

▶ **Firn Field Pressure** The firn field pressure results from the pressure exerted by the glacier ice mass in the nourishing area on the ice masses in front of it.

The flow velocity of a glacier depends on various factors: the gradient at the base of the glacier, the pressure of the firn field, the nature of the bedrock and the cross-section of the flowing ice. If the cross-section narrows, the ice flows faster. The flow velocity within a glacier is not the same everywhere. It increases above the firn line and decreases below it. It is greater in the middle of the glacier than at the edge, and it

Fig. 2.6 Warning of breaking ice in front of the glacier train of Bear Glacier near Seward (Alaska): "Do Not Go Beyond This Point" – "Do not enter the area beyond the barrier". (© Wolfgang Fraedrich)

is greater in the upper section of the glacier at its base than in the lower section, where it is greater at the surface. Some alpine glaciers flow up to 200 m, some Greenland glaciers even up to 7 km per year.

In Alaska, too, numerous glaciers have high flow rates. Warning signs have been erected everywhere prohibiting approaching the immediate ice front of a glacier tongue (Fig. 2.6). Due to the advance of the ice, suddenly breaking off blocks of ice can be life-threatening, as tourists learn painfully year after year when they defy the warnings.

However, this high flow velocity must not be equated with a corresponding increase in the glacier's size, as the glacier also loses mass, especially in the ice flow area. The type of ice movement is related to the different flow velocities, but also to the temperature of the ice. The ice of temperate glaciers is just below freezing point. Slow flowing temperate glaciers have a streaming motion, they flow like a viscous mass (like a pudding). The faster flowing cold glaciers with comparable relief conditions move like elastic blocks. One speaks then also of a block floe or block movement.

As the glacier moves, its appearance also changes. Glacial crevasses form as a result of the flow process. Transverse crevasses form on the upper side of the glacier, especially above (convex) terrain steps, and on its lower side where (concave) depressions are flowed through. Longitudinal crevasses occur when the glacier tongue widens. Due to the faster movement of the ice in the centre of the glacier, marginal crevasses form in the temperate, slow-flowing glaciers. The somewhat slower flowing glacier margin is divided into a multitude of individual blocks, called **séracs**, which also form when a glacier flows downhill over a steep slope. The fact that crevasses and séracs form at all can be explained by the fact that the ice does not react as plastically as the respective (surface) stress requires. In the case of block movement, on the other hand, the ice velocity is almost the same throughout the entire cross-section.

2.7   Glaciers Are Growing and Melting

Glaciers advance and they "recede", even a "receding" glacier flows forward, as paradoxical as that may sound. If it loses mass, it nevertheless continues to flow, only its ice edge shifts backwards. The advance results from a positive mass balance. If the mass balance is negative, the ice edge moves more and more backwards.

The **mass balance** of the glacier is essentially shaped by two climatic factors. Firstly, the amount of precipitation in the nutrient zone. If this is low, not enough ice can form, the firn field pressure is low, and the mass balance is negative. The same applies, of course, to decreasing precipitation over a longer period of time. Secondly, by the temperatures. If these fall, less ice will melt during the warmer months; if temperatures rise, a greater amount of ice will inevitably be lost through melting processes.

▶ **Mass Balance** The mass balance of a glacier determines whether a glacier grows or loses volume. A positive mass balance is given when less ice is lost through melting processes than is added to the nutrient area by a corresponding amount of new snow in the same period.

2.8   Glacial Fluctuations and Their Causes

Changes in the mass balance of a glacier thus cause a glacier to increase or decrease in size. This is also referred to as a change in the glacier balance or the ice balance. Changes in the glacier balance can be determined by taking different time periods into account.

The measured values refer to:

- the period of 1 year,
- the average of several years or even
- a very long period of time.

It has also already been shown that changes in the mass balance become apparent in the advance or retreat of the ice margin. The changes in a glacier can best be observed in valley glaciers. As a rule, the length fluctuations are many times greater than the width fluctuations. Even with small ice streams the ratio is 10:1, with long valley glaciers it reaches a ratio of >1000:1. Even more insignificant than the glacier width fluctuations are the level fluctuations. These document the overall spatial function of the glacier budget, which is in any case very difficult to determine, even if a glacier is surveyed at several points using gauge observations. This also explains why one generally speaks of glacier advance or retreat and less of the volume change of a glacier.

The annual glacier fluctuations are primarily weather-related, they are a result of the constant interplay of snowfall and melting (= **ablation**). Already the daily, but also the multi-day and finally also the seasonal weather changes play the main role.

▶ **Ablation** Ablation refers to the melting and evaporation of glaciers as well as snow masses – especially on their surface – caused by solar radiation, back radiation from rock, atmospheric heat, air humidity, air movement and surface run-off of melt and precipitation water.

In contrast to weather-related fluctuations, the mass changes of a glacier observed over the longer term are the result of **climate fluctuations** or even **climate (change)**.

Climate Fluctuations Climate fluctuations are clearly measurable changes in the course of the climate with not too great an impact, which also do not last too long. In historical times, for example in the nineteenth century, there was a significantly colder phase during which the glaciers advanced considerably. At the end of the twentieth century and currently at the beginning of the twenty-first century, we are living in a warmer phase, which is expressed worldwide by glacier retreat. A global temperature change of only 0.5 °C is considered a climate fluctuation. Climate fluctuations occur both during warm periods and during cold periods (Fig. 2.7).

▶ **Climate (Changes)** Climate (changes), on the other hand, are much more far-reaching. Temperature changes reach higher values (about 5 °C and more) and can lead to fundamental climate change (e.g. the transition from a cold to a warm period, cf. Fig. 2.7).

Changes in the glacier budget are therefore weather- and/or climate-related. Changes in weather patterns can be explained by numerous basic meteorological principles, which in turn are controlled in their varying relevance by the seasonally varying radiation budget of the Earth.

The long-term climate fluctuations and climate (changes) are also the result of a changed radiation budget. However, this has causes that are much more complex to consider and will therefore not be discussed in detail at this point. The final Chap. 9 will present the current state of research in this area.

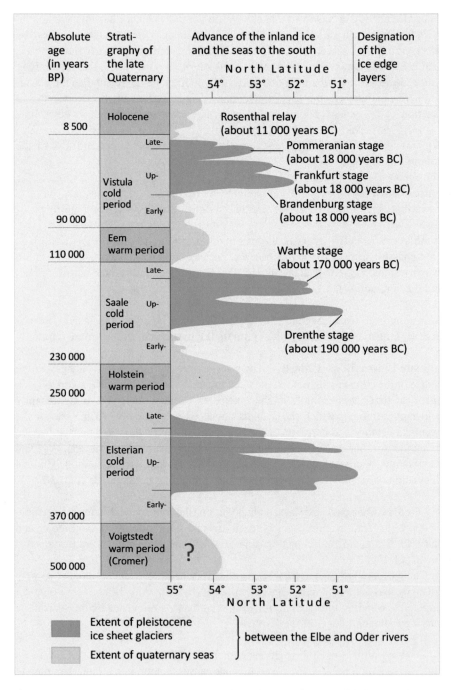

Fig. 2.7 Advance of inland ice glaciers and sea-level rise over the last 500,000 years of Earth's history for the area of northern and central Europe as an expression of climate variability and (changes in) climate. (© Wolfgang Fraedrich)

2.9   The Registration of Glacier Fluctuations

Skaftafellsjökull in southern Iceland, a valley glacier at the edge of the Vatnajökull ice sheet, is retreating. The deposits of the last advance phase before the ice edge are a visible document of its retreat (Fig. 2.8). It must therefore have become warmer, otherwise the glacier margin would not have retreated. In what period did the glacier retreat take place?

Such developments have taken place throughout Europe, in Greenland and also in North America. In numerous studies of glaciers and through the evaluation of meteorological and glaciological records of the recent past, scientists have found that a "**Little Ice Age**" took place in the nineteenth century. In scientific literature, the term "**Little Ice Age**" has become established.

It is not least the concern about human-induced warming of the Earth's atmosphere that is increasingly prompting scientists to observe glaciers. Changes in the glacier budget allow conclusions to be drawn about possible climate fluctuations, and studies of glacier ice provide information about the climate in the past and create the preconditions for forecasts about the climate of the future (cf. Chap. 9).

Glaciers have been observed for more than 100 years, and their changes are being recorded with increasing detail and precision. Scientists are working in all glacier regions of the world in order to be able to draw conclusions for future climate development from the results obtained. The comparison of the model representations

Fig. 2.8 Glacier forefield of Skaftafellsjökull in southern Iceland. (© Wolfgang Fraedrich)

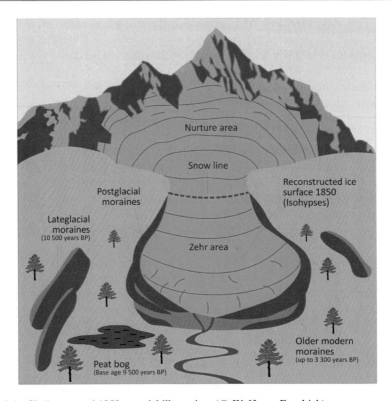

Fig. 2.9 Glaciers around 1850 – model illustration. (© Wolfgang Fraedrich)

in Figs. 2.9 and 2.10 shows the glacier retreat since the peak of the "Little Ice Age" around 1850. The evaluation of the glacier inventory of 1850 and the subsequent measurements and investigations in comparison with the present state of the glaciers have revealed enormous mass balance changes in the eastern Swiss Alps, for example.

Although there were short phases of advance in the meantime, around 1890 and 1920, the glaciers as a whole have clearly lost length, area and thus also volume (Table 2.2).

The mass loss of Alpine glaciers has also continued in recent times. While the Pasterze at the foot of the Großglockner massif lost about 24 km^2 in area between 1850 and 2004 (an area loss of about 20%), significant losses were also recorded in 2010 and 2011. In 2011, the end of the glacier tongue of the Pasterze receded by 40.3 m (2010: −24.7 m), while at the same time the glacier tongue lost an average of 4.4 m in thickness (2010: −1.4 m). Figure 2.10 – taken in August 2019 – shows how far the glacier front has receded since 2010 – in only 9 years (!) (Fig. 2.11).

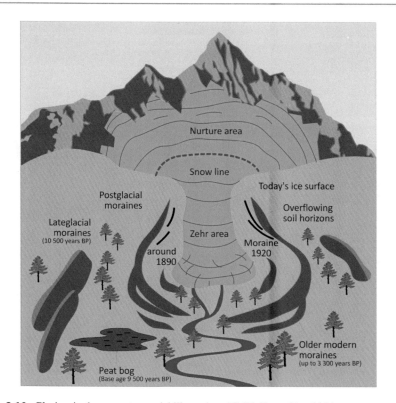

Fig. 2.10 Glaciers in the present – model illustration. (© Wolfgang Fraedrich)

Table 2.2 Glacier recession of the Morteratsch glacier (eastern Switzerland)

	12,300 BP	1850	2000	Loss 1850–2000 (%)
Length (km)	16.0	9.0	7.0	−22.2
Area (km²)	32.0	19.3	16.4	−15.0
Ice thickness	95	75	70	−6.7
Volume (km³)	4.0	1.5	1.2	−20.0

Fig. 2.11 Marking of glacier ice margins as evidence of glacier retreat, in a period of only 9 years the glacier has melted back a long way. (© Wolfgang Fraedrich)

Distribution of Ice and Cold Regions

3

Wolfgang Fraedrich

Abstract

In order to be able to assess current developments – also in the course of global warming – more precisely, an overview of the spatial dimensions of the areas on Earth shaped by ice and ice climate is helpful and important. This chapter provides an overview of

- the spatial dimensions of glaciated regions and permafrost areas on Earth,
- the differences between glacial ice, ice shelves and pack ice,
- Trends in changes in ice extent and permafrost regions.

A "virtual excursion" around the globe with Google Earth® is an information opportunity to learn about areas on Earth covered by ice. With a total area of about 14.9 million km², glaciers cover about 10% of the Earth's land surface. In addition, the earth's surface is covered with sea ice, about 7% of the ocean surface is iced. In addition to the icy regions, the zone of permafrost is closely linked to the glacier regions, covering about 20 million km². Similar climatic conditions prevail here, there are also very cold winters and – depending on latitude or altitude – only short, relatively cool summers (subpolar or subarctic climate).

If you want to get an idea of the spread of glaciated areas on Earth, a "virtual excursion" around the globe with Google Earth® is one way of obtaining information. Depending on the radiation budget and thus on the prevailing air temperature, glacier ice spreads primarily in the higher latitudes and here primarily in the polar regions. In comparison, the glacier areas in the high mountain ranges of the earth

W. Fraedrich (✉)
Hamburg, Hamburg, Hamburg, Germany
e-mail: wolfgang.fraedrich@hamburg.de

W. Fraedrich, *Traces of the Ice Age*, https://doi.org/10.1007/978-3-662-65886-4_3

occupy only very small areas. With a total area of about 14.9 million km², glaciers cover about 10% of the earth's continental surface, whereby the continent of Antarctica is covered by glacier ice on an area of 12.5 million km². Greenland has a glacier area of about 1.7 million km². Only about 700,000 km² (which is less than 4% of the total glaciated area of the Earth) are covered by glaciers in the remaining polar areas as well as the high mountains of the Earth. In addition, the Earth's surface is covered with sea ice; about 7% of the ocean surface is iced over.

In addition to the icy regions, the zone of **permafrost**, i.e. the region of permanently frozen ground, is closely related to the glacier regions. Similar climatic conditions prevail here, there are also very cold winters and – depending on latitude or altitude – only short, relatively cool summers. Climatologists speak – in distinction to the polar climate – of the subpolar (in the northern hemisphere of the subarctic) climate. About 20 million km² of the continental area (just under 14%) are affected by subpolar climatic influences and are therefore permafrost areas.

Humans have long since penetrated these regions and also live there in permanent settlements. However, there are considerable limits to land use, because the short summers only allow for a very short vegetation period and because, in addition, the ground only thaws at the surface for a short time and only to shallow depths. Even the construction of houses and transport routes, or even oil or gas pipelines, as in Alaska or northern Siberia, poses numerous problems because the thawed ground does not provide sufficient stability as a subsoil.

The change from summer thawing to winter freezing is, moreover, the trigger for numerous soil and relief altering processes, which are explained with in more detail as so-called periglacial formation processes in Chap. 7.

At about 35 million km², the regions of perpetual ice and permafrost cover almost a quarter of the Earth's total land mass.

The Earth's ice area is still increased by about 26 million km², which corresponds to about 7% of the world's ocean area. However, it should be borne in mind that this is an annual average, as sea ice cover is subject to seasonal fluctuations that have different peaks. While in the Arctic Ocean the greatest sea ice extent is reached in March, in the Antarctic region the maximum values are reached in October. The glaciation of the polar oceans extends over the whole year, but only in winter a closed ice cover of >4 m thickness is reached. This closed ice cover thaws only at a thin surface layer, which soon breaks apart at crevasses, creating drifting ice islands. The transition from the inland ice of the continental regions (e.g. Greenland and Antarctica) takes place via the **ice shelves** and the pack ice to the sea regions that are influenced by drifting icebergs, which cover a further 64 million km². The total area influenced by sea ice thus covers about a quarter of the world's ocean surface.

▶ **Ice Shelf** Ice shelves are defined as the immobile mass of ice lying on a shelf and consisting of continuous ice sheet or a mixture of sea ice and broken ice sheet.

▶ **Pack Ice** Pack ice is understood to be the ice floes pushed or pressed together into large ice masses essentially by wind pressure.

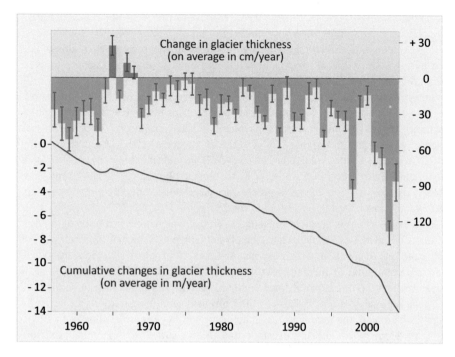

Fig. 3.1 Glacier retreat worldwide (© Wolfgang Fraedrich)

However, the largest amount of terrestrial sea ice is not formed in the sea, but on the mainland; it therefore also stores fresh water. With his "ice shelf hypothesis" presented in 1964, the American A. T. Wilson scientifically substantiated the idea according to which inland ice can only reach a so-called maximum thickness. If this is exceeded, pressure melting begins on the underside, which initiates the slipping of the ice into the sea. If ice masses detach themselves from the so-called "parent glacier", this is called calving of a glacier, and icebergs are formed in the process. According to estimates, icebergs with a volume of almost 1500 km^3 are formed annually in the Southern Ocean alone, which corresponds to about half of the world's annual freshwater demand.

Due to the warming of the Earth – this phenomenon is addressed more intensively in Chap. 9 – glaciers and also sea ice areas are in retreat worldwide. In the last 20 years alone, the global glacier ice area has decreased by 1.1 million km^2 (= 6.9%), the thickness of the glaciers is also declining (Fig. 3.1). Among other things, this is leading to significant changes in the water balance of adjacent regions and ultimately to changes in ecosystems. Sea ice areas are also declining, both in terms of area and ice thickness – it is expected that the Arctic Ocean will be ice-free in summer as early as 2040. "The extent of Arctic sea ice reached its all-time minimum observed in September 2012. (. . .) The thickness of sea ice has decreased significantly. The most common sea ice thickness during summer was about 3.0 m in the 1960s, still over 2.0 m in the 1990s, and about 0.9 m in the last two years" (Nicolaus and Hendricks 2015).

Rising temperatures will also lead to changes in the permafrost regions. The permafrost soils will lose thickness, while the thaw layers will become thicker. Depending on the relief, this will result in morphological changes (cf. Chap. 7). A major problem will also be the release of greenhouse gases bound in the frozen soil, because as the soil thaws bacteria and microorganisms become active and begin to decompose animal and plant remains bound in the permafrost. These processes release mainly carbon dioxide (CO_2) and also methane (CH_4). According to the Alfred Wegener Institute in Bremerhaven (AWI), an additional increase in the global mean temperature of almost 0.3 °C is to be expected by the end of the twenty-first century as a result of the release of greenhouse gases due to ongoing global warming, and even by more than 0.4 °C by the year 2300.

Another problem is posed by methane hydrates, which are a frozen compound of methane and water molecules. Here, the crystal lattice of the water molecules forms the "cage", so to speak, in which the gas (methane) is "trapped". Methane hydrate is a frozen compound of methane and water molecules, the latter forming a kind of cage in which the gas molecules are enclosed (chemically unbound). In continental subpolar regions, methane hydrates – their physical stability conditions are given by high pressure and low temperatures – can occur within permafrost soils from 100 to 2000 m depth below the land surface due to low surface temperatures and low temperature gradient.

Glaciations During the Geological Past

4

Wolfgang Fraedrich

Abstract

The Earth's climate has never been constant. Both, extraterrestrial and terrestrial processes, contribute to the fact that the earth's climate is the result of extremely complex dynamic processes in a constantly changing system. However, this need not be grasped in all its complexity. Rather, it is a matter of grasping basic interrelationships and learning to place them in the overarching context. In this chapter, the you will learn,

- that the earth's climate is subject to constant change,
- that the current ice age is in principle a repetition of geological epochs that have already shaped the history of the earth over millions of years,
- in which temporal dimensions and which spatial dimensions the current ice age is to be classified,
- which stratigraphy is characteristic for the recent development of the earth,
- how geologists reconstruct climate history,
- how the cold ages have affected Europe.

In Chaps. 2 and 3, the main principles for understanding the formation of relief caused by ice have been explained. Occasional reference has already been made to processes in the recent geological past or in historical times (cf. Chap. 2). In particular, observations from historical times, but also observations that scientists can make in the present, often allow conclusions to be drawn about events in the geological past. This is a central principle of geological research, the so-called **actualism principle**.

W. Fraedrich (✉)
Hamburg, Hamburg, Hamburg, Germany
e-mail: wolfgang.fraedrich@hamburg.de

© The Author(s), under exclusive license to Springer-Verlag GmbH, DE, part of Springer Nature 2023
W. Fraedrich, *Traces of the Ice Age*, https://doi.org/10.1007/978-3-662-65886-4_4

In particular, observations from historical times, but also the observations that scientists can make in the present, often allow conclusions to be drawn about events in the geological past. This is a central principle of geological research, the so-called actualism principle. In relation to glacial research, this means, for example, that observations of glaciers that can be made in the present are transferred to processes in prehistoric times. This applies not only to the most recent geological periods of the Quaternary, but also to proven glaciations of more distant periods. In this context, it is important to realize that throughout the entire history of the Earth there have always been varying climatic conditions of very different temporal scales.

▶ **Actualism Principle** The principle of actualism is based on the assumption that forces, processes and phenomena of the geological past are similar to those observed in the present. This makes it possible to draw direct conclusions from the appearance or forces and processes observed in the present to earlier ones. Even though temporary peculiarities in the development of the earth's history have been demonstrated, the principle of actualism is now generally accepted.

In terms of glacial research, this means that, for example, observations of glaciers that can be made in the present are transferred to processes in prehistoric times.

This applies not only to the most recent geological periods of the Quaternary, but also to proven glaciations of more distant periods.

It is important to bear in mind that throughout the Earth's history (Table 4.1) there have always been varying climatic conditions of very different temporal scales. Geologists distinguish between

• long-term climate changes of several 10 million to 100 million years,
• medium-term climate changes of several 10,000 to 100,000 years and
• short-term climate fluctuations on the order of a few 100 or 1000 years.

Only a little more than one seventh of the earth's climatic history, corresponding to the period since the beginning of the Cambrian 541 million years ago, can be reconstructed in more detail.

The numerous investigations since the beginning of geological research have yielded some information about the climate in the Neoproterozoic, but numerous traces that would have allowed an exact conclusion about the climatic conditions of that time have been removed, i.e. "obliterated", so to speak, by the dynamic processes in and on the earth – among others rock processing (= weathering), ablation (= erosion) and sedimentation (= accumulation) or also volcanism and plate tectonics.

The Snowball Earth hypothesis, put forward in 1992 by the US geologist J. L. Kirschvink and still controversial among experts, is based on the assumption that the Earth was glaciated from the poles to near the equator and that the oceans were also largely frozen. Kirschvink assumes that this phase covered the period 750–580 million years before today in the younger Neoproterozoic. Glacial deposits corresponding to this phase, known as tillites, have been found on all continents. The fact that these tillites occur in association with rocks that are more likely to be

Table 4.1 Earth history time table (compared to a calendar year)

Begin[a] (Both is correct million years BP)	Duration (million years)	Formation	Evolutionary stages of animal life	Calendar date
2.6	2.6	Quaternary	First people	31.12. (19:45)
66	63.4	Tertiary	Evolution and dispersal of mammals	26.12.
145	79	Chalk	Towards the end, the dinosaurs become extinct, the first mammals and bony fish	20.12.
201	56	Jura	The heyday of the dinosaurs, ammonites, belemnites, crinoids	16.12.
252	51	Triad	Dinosaurs rule the land	13.12.
299	47	Permian	First ammonites	09.12.
359	60	Carbon	First reptiles, the seas are dominated by fishes	04.12.
416	57	Devon	First insects, transitional forms of fish and amphibians	29.11.
444	25	Silurian	First vertebrates	25.11.
488	41	Ordovician	Increase in the number of marine invertebrate species	20.11.
542	57	Cambrian	First marine invertebrates	15.11.
4600	4059	Precambrian	First algae in the primal ocean	01.01.

[a]Annualanga ben based on the *International Chronostratigraphic Chart of* the *International Commission of Stratigraphy.* (cf. http://www.agiweb.org/nacsn/67209_articles_article_file_1639.pdf)

found in tropical regions led to the conclusion that glaciation must also have been possible near the equator. Plate tectonic processes, due to which precipitation conditions changed, especially in the area of the continents, and as a consequence partly significant changes of the CO_2 content in the atmosphere were probably responsible for the initially significant decrease of the global mean temperature. Towards the end of this glaciation period, changes in the arrangement of the continents and a higher concentration of CO_2 in the atmosphere, among other things due to stronger volcanism, were also responsible for the increase in the global mean temperature.

During this snowball period, as in today's cold ages, there were probably cold and warm periods. Thus, it is thought that there were at least two more significant cold periods, the Sturtian Cold Period (715–710 million years B.P., named after South Australia's Sturt Valley, where tillites have been found) and the Marinoan Cold Period (named after the small town of Marino in South Australia), which spanned the last million years of the outgoing Cyrogenian, which ended 630 million years ago.

4.1   Pre-quaternary Cold Periods

As for every climate, there is also evidence for cold-age climates. Finds of
pre-quaternary glacial traces are, for example, glacier scars and solidified ground
moraine material, so-called **tillites**.

▶ **Tillit** In German, tillite is the term used to describe ground moraine remnants of
pre-Quaternary cold periods that have been consolidated by pressure and
temperature.

Such tillites and glacial scars have been found on various continents (Fig. 4.1),
thus essentially three major periods of cold climate have been demonstrated
(Fig. 4.2): the Quaternary, the Permo-Carboniferous glaciation at the Carbonifer-
ous/Permian boundary, and the so-called Eocambrian glaciation in the late
Neoproterozoic.

The original attempt to deduce a periodicity in the occurrence of cold-temperate
climatic conditions, i.e. an alternation of cold and warm periods, from the temporal
distribution of the finds has been refuted by the finds of the tillites.

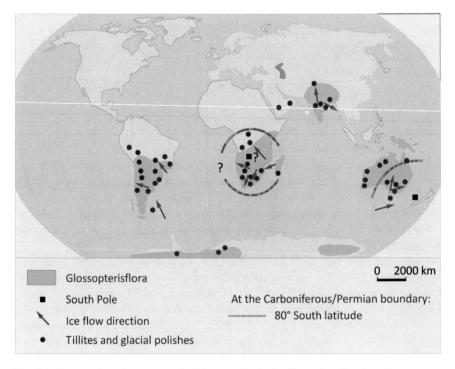

Glossopterisflora

■   South Pole

\   Ice flow direction

•   Tillites and glacial polishes

0   2000 km

At the Carboniferous/Permian boundary:
.... 80° South latitude

Fig. 4.1 Traces of pre-Quaternary glaciations at the Carboniferous/Permian boundary, approx.
280 million years ago (© Wolfgang Fraedrich)

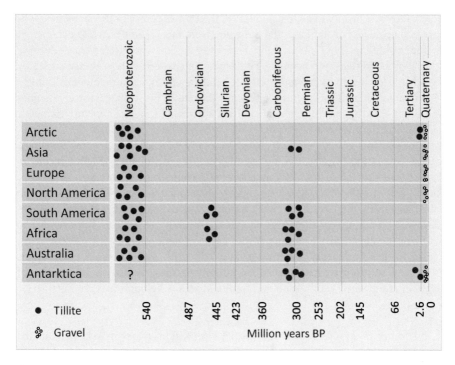

Fig. 4.2 Proven glaciations since the beginning of the Cambrian, about 542 million years ago (© Wolfgang Fraedrich)

4.2 Quaternary Cold Periods

The most recent period of the Earth's history, which geologists call the Quaternary, is characterized by a constant change of colder and warmer climates. From a climatic point of view, the Quaternary is divided into the Pleistocene (2588 million years to 11,700 years BP) and the Holocene (present warm period, since 11,700 years ago). After the glaciation phases of the Late Cambrian and the Late Palaeozoic, the third major glaciation phase of the last 540 million years began in the Quaternary. According to the current state of knowledge, it can be assumed that the Holocene is also only an intermediate warm period, i.e. that another cold period will follow in the geologically near future. The background to this assumption is discussed in more detail in Chap. 9.

4.3 The Keys to the Past

Geologists have gained a fairly precise idea of the course of the Quaternary evolutionary history. In contrast to the past cold periods, there are a large number of "traces", especially in the northern hemisphere, but also in the southern hemisphere, which allow the history to be traced back. In the further course of this book they will be traced again and again. First, the transition from the Tertiary to the Quaternary is the focus of interest.

The temporal transition from the Tertiary to the Quaternary was characterized by a gradual decrease in global temperature. While a subtropical to temperate climate prevailed in today's Central Europe towards the end of the Tertiary, the Quaternary was dominated by the repeated alternation of subarctic and temperate climates, especially in the northern hemisphere.

However, this gradual climate change can only be traced at very few localities. One of these points of particular interest to geologists is Meinert's "Lieth lime pit" in the northwest of the district of Pinneberg near Hamburg (coordinates: N 53° 43′ 15.27″, E 9° 40′48.49″), which was already abandoned in the 1980s. In connection with the rise of the Zechstein salts, which is typical for large parts of northern Germany, numerous karst quarries occurred, especially in the late Tertiary, due to the intensive chemical weathering of the easily soluble sediments (= **evaporites**) typical for the **Zechstein – including** lime, anhydrite and gypsum. Cave systems formed by **solution weathering** collapsed, and in these sinkholes a sequence of layers of redeposited kaolin sand and **lignite seams developed in** a confined space (Figs. 4.3 and 4.4).

▶ **Evaporite** Evaporites represent a special form within sediments. They are not the result of an accumulation of mechanically weathered rock debris (so-called clastic rock debris), but are formed by the evaporation of water (especially seawater) under dry-hot climates. The ions dissolved in the water are precipitated in the process, since water cannot contain an unlimited amount of dissolved substances. According to their degree of solubility, carbonates (e.g. limestone) are formed first, then sulphates (e.g. anhydrite) and finally the most soluble chlorides (common salt and potassium salt).

▶ **Zechstein** During the Zechstein, the north of today's Central Europe was repeatedly covered by shallow seas, which at times even had no exchange of water with the ocean. Under the arid climatic conditions of the Late Permian, millions of cubic metres of seawater evaporated over a very long period of time. What remained were thick sedimentary layers which, due to their plasticity, later rose in many places in the North German Lowlands and in the North and Baltic Sea basins, and in some cases – as in the case of the Elmshorn salt dome near Pinneberg – even reached the surface.

While sands were initially blown into the sinkhole, a "small hole" in the earth's surface at that time, at the beginning of the Cold Ages, a warmer climate in the meantime allowed soil to form on it, which in turn made plant growth possible. When it became colder again for centuries, the plants gradually died and they were covered again by sand, which could be blown out of the vegetation-poor landscape elsewhere and into the sinkhole. With a renewed rise in temperature, soil formed

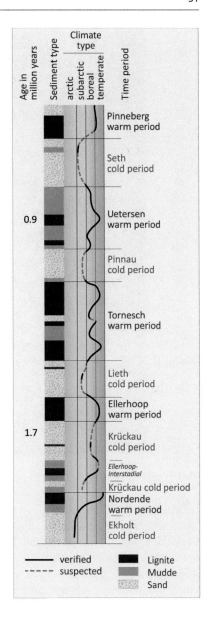

Fig. 4.3 Stratigraphy of the Older Quaternary–"Lieth Series" (© Wolfgang Fraedrich)

again, plants settled, which later died again, and so on. Due to the growing load, the lower layers were compressed and thus an incarbonation process began.

The Nordic inland ice, which penetrated north-central Europe several times during the later glaciation periods, repeatedly covered this small area of less than 1 ha for a longer period of time, but was unable to destroy the original layered structure. Like the filling of a tooth, it was able to survive undisturbed until the present day; it was only examined for the first time in 1940. In the northwest of the

Fig. 4.4 Outcrop of the so-called "Lieth Series" (© Wolfgang Fraedrich)

pit, which at that time had already existed for decades and in which the various Permian rocks had been quarried, first the sands were discovered and later also interbedded lignite seams. In the following three decades numerous scientific investigations took place. First the sands and only later the lignite seams found increased interest. The age of the kaolin sands and also of the lignite could be determined quite precisely by **C14 dating** of the remains of dead organic substances. In addition, **pollen analyses** of the individual lignite seams also provided a fairly accurate picture of the vegetation communities of the time.

Thus, not only was it clear that the sands were formed in cold phases and the lignites in warm phases, but on the basis of the pollen analysis results it was even possible to determine that the first cold phases in the Quaternary were by no means to be equated with the most recent cold phases from an ecological point of view, but rather marked the transition from a phase of persistently warm climate in the late Tertiary to a climate characterised by cold phases.

▶ **C14 Dating** In the course of C14 dating, carbon is extracted from fossil organic material and the content of a radioactive carbon isotope with the mass number 14 (instead of the mass number 12 for "normal" carbon) is determined with the aid of complicated and extremely sensitive measuring equipment. Hence the name C14 method (or radiocarbon method). This method is very accurate because in the uppermost layers of our atmosphere the C14 content in the Earth's atmosphere has been kept constant for thousands of years. There is about 10^{-1}‰ C14 in atmospheric carbon dioxide. Since all living plants also constantly absorb C14 through respiration and since these are also the food basis for animals and humans, all living beings have a share in this C14. After death, however, this begins to decay, while the carbon with the mass number 12 remains unchanged. The half-life of C14, i.e. the time in which half of the carbon present decays, is 5730 years. Accordingly, after 11,460 only a quarter, after 17,190 years only an eighth of the original C14 is still present, and so on. These findings allow a quite exact determination, which in the most extreme case can go up to 70,000 years into the past (then only about 0.025% of the originally present C14 can be measured).

▶ **Pollen Analysis** The pollen count of the present day determines the everyday life of almost 50% of people in highly industrialized Central Europe at different times of the growing season, which begins in spring and ends in autumn. Microscopically tiny pollen grains and spores are triggers of mucosal allergies.

In science, they are very important indicators for determining the prehistoric climate within the framework of the so-called pollen analysis. Many pollen and spores can survive for millennia without ever being completely decomposed. While the pollen content dies quickly, the outer pollen wall can be preserved if it is exposed to air as soon as possible.

For the analysis of peat, mud or clay samples only about 1–3 cm^3 are needed, impurities should be avoided. Since the pollen grains and spores contained in the sample are to be extracted for analysis, the sample is chemically treated in several steps in the laboratory in order to extract as much as possible of the remaining substances. Various acids (e.g. hydrofluoric acid), some of which are highly corrosive, are used to destroy quartz, lime and cellulose, among other things. If necessary, the pollen is bleached or dyed, then viewed under a microscope (a 1000-fold magnification is ideal) and finally compared with an identification key and counted.

4.4 Pleistocene (Cold Ages)

If one wants to get an overview of the entire developmental history of the Quaternary, it makes sense to focus more and more on the path from the geological past to the present. Due to the well-preserved structures, one is able to develop quite precise ideas, especially for the most recent **cold** and warm periods. The **stratigraphic tables** in Figs. 4.5, 4.6 and 4.7 serve as a constant orientation.

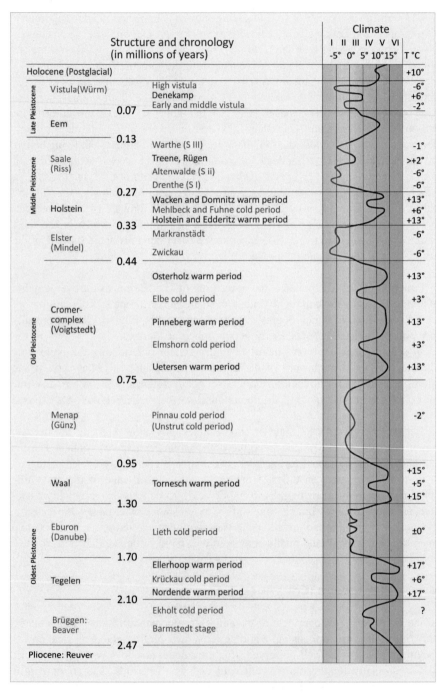

Fig. 4.5 Breakdown of the Quaternary period (© Wolfgang Fraedrich)
The climate curve shows the annual mean temperatures (T in °C); the Roman numerals mean:
I = frost tundra, II = shrub tundra, III = birch-pine forest, IV = cool temperate forest vegetation,
V = warm temperate forest vegetation, VI = Mediterranean vegetation

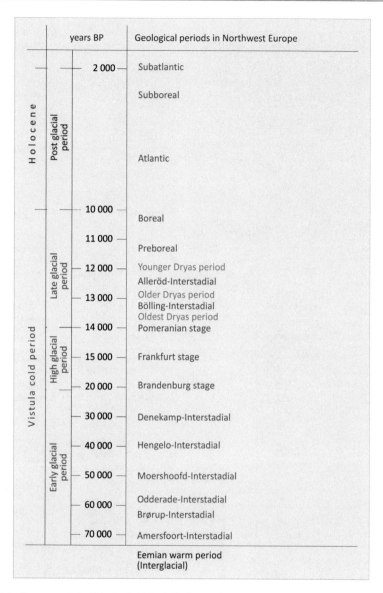

Fig. 4.6 Structure of the Vistula Cold Period (© Wolfgang Fraedrich)

▶ **Cold Period** As a technical term, "cold period" is more accurate than the word "ice age", because it is not only the dominance of ice in certain regions of the earth that has a climate-determining and relief-shaping effect, but because clearly perceptible influences are also detectable – and in the present also directly ascertainable – away from the directly icy regions.

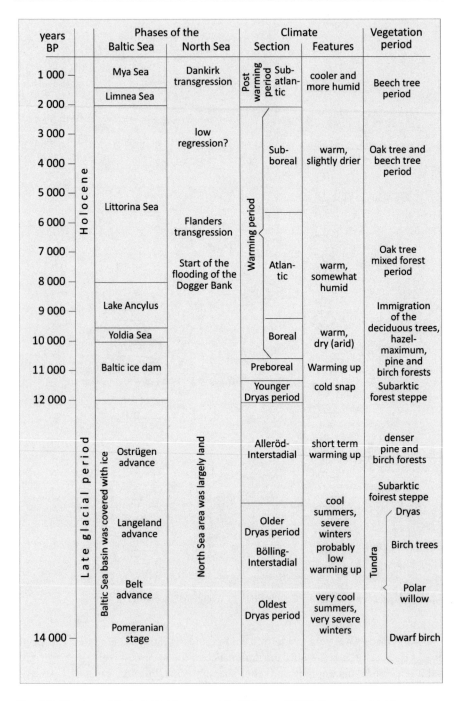

years BP	Phases of the Baltic Sea	Phases of the North Sea	Climate Section	Climate Features	Vegetation period
1 000	Mya Sea	Dankirk transgression	Post warming period / Sub-atlantic	cooler and more humid	Beech tree period
2 000	Limnea Sea				
3 000		low regression?	Sub-boreal	warm, slightly drier	Oak tree and beech tree period
4 000					
5 000	Littorina Sea				
6 000		Flanders transgression			Oak tree mixed forest period
7 000			Atlantic	warm, somewhat humid	
8 000		Start of the flooding of the Dogger Bank			
9 000	Lake Ancylus				Immigration of the deciduous trees, hazel-maximum, pine and birch forests
10 000	Yoldia Sea		Boreal	warm, dry (arid)	
11 000	Baltic ice dam		Preboreal	Warming up	
12 000			Younger Dryas period	cold snap	Subarktic forest steppe
	Ostrügen advance		Alleröd-Interstadial	short term warming up	denser pine and birch forests
					Subarktic foirest steppe
				cool summers, severe winters	Dryas
	Langeland advance		Older Dryas period		
			Bölling-Interstadial	probably low warming up	Birch trees
	Belt advance		Oldest Dryas period	very cool summers, very severe winters	Polar willow
14 000	Pomeranian stage				Dwarf birch

Holocene (from 1 000 to ~11 500 BP), Late glacial period (lower section). Baltic Sea basin was covered with ice. North Sea area was largely land. Warming period. Tundra.

Fig. 4.7 Structure of the Late Pleistocene and the Holocene (© Wolfgang Fraedrich)

▶ **Stratigraphy** Stratigraphy is a branch of geology that orders rocks according to their chronological order of formation, taking into account all their organic and inorganic characteristics and contents, and establishes a time scale for dating geological processes and events.

4.5 Pleistocene Europe

During the Quaternary cold periods, large parts of the high latitudes and all of the Earth's high mountains were repeatedly covered by extensive inland ice and glacial ice masses. This is true for North America, North Asia, South America and for Northern Europe as well as parts of Central Europe.

According to the current state of geoscientific research, the glaciations of Central Europe (e.g. in the Alpine region) and Northern Europe are largely clear. It is assumed that during the earlier cold periods glaciations occurred only regionally limited. These are likely to have been spatially arranged in a similar way as is the case in the present, albeit with a greater extent. According to the respective regions, they have been given region-specific names derived from rivers in northern Germany (Elbe, Elster, Saale, Weichsel) and in southern Germany (Günz, Mindel, Riss, Würm).

For northern and north-central Europe, the first significant inland ice advance is assigned to the Elster cold period. The ice reached the northern edge of the low mountain range zone and even advanced around the eastern edge of the Harz Mountains into the western part of the Thuringian Basin. Almost the entire area covered by ice during this time was glaciated again in the course of the Saale and Riss Cold Ages. While in the foothills of the Alps no sediments of the Mindel (= Elster) glaciation are found on the surface, corresponding finds in north-central Europe are limited to Thuringia and southern Saxony. Due to the erosion processes, the corresponding forms have long since ceased to be detectable. Also the ice margins of the Saale and Risskalk periods – if they have not been overrun by the inland ice advancing again during the Weichselian and Würmkalk periods – have been strongly flattened by erosion processes in the course of the last 120,000 years. The most conspicuous surface formations extend south of the Lower Elbe in the area of the Lüneburg Heath from Hamburg (Harburg Mountains) and east of Lüneburg (Drawehn) towards the southeast – this is the ice margin of the Warthe Stage – and almost from west to east through central Lower Saxony to Saxony-Anhalt (roughly from Lingen on the Ems to Magdeburg) – this is the ice margin of the Rehburg Stage. For the Eemian warm period (approximately from 110,000 to 80,000 before today) similar climatic conditions are assumed as they are typical for the present. Investigations in the Baltic Sea region have shown that Scandinavia was probably an island complex during this time (Fig. 4.8). During the Weichselian or Würm glacial period, large-scale glaciation occurred again and was relieved in several advances (Fig. 4.9). It was not until about 12,000 years ago that the ice finally began to retreat.

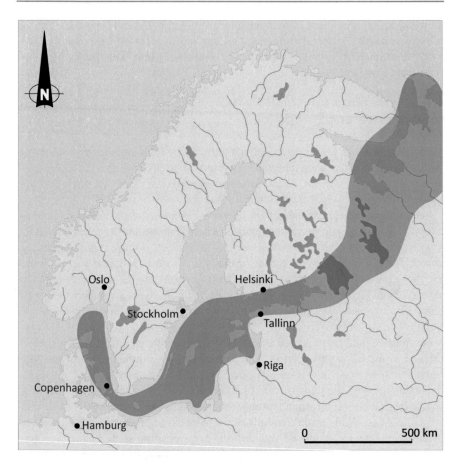

Fig. 4.8 The so-called Portlandia Sea of the Eemian Warm Period in the area of today's Baltic Sea (©Wolfgang Fraedrich)

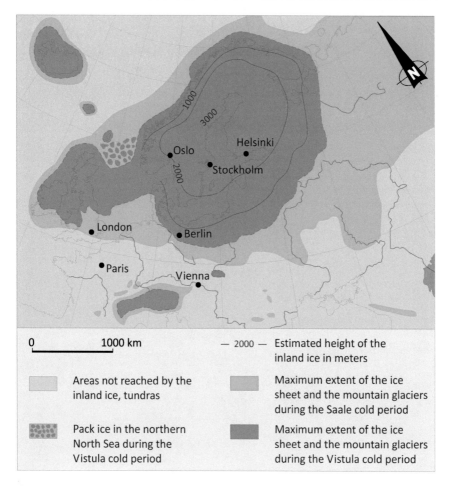

Fig. 4.9 Distribution of ice in Europe during the last cold periods (© Wolfgang Fraedrich)

Glaciers Shape Landscapes

5

Wolfgang Fraedrich

Abstract

Glacial ice can flow. And flowing ice is a force that leaves clear traces in the landscape through various processes such as ablation and deposition. These become visible above all after melting.

In this chapter, the reader learns both the basics and some important details on the subject of "glacial forms", in detail

- it is explained where exactly the difference between different relief-forming forces and processes lies,
- the glacial depositional forms are presented in an overview,
- the importance of lead-grains for the analysis of glacial regions is explained,
- there are "instructions for action" for own investigations at moraine outcrops,
- the moraines are presented in detail as particularly prominent depositional forms, distinguishing between north-central Europe and mountainous areas,
- numerous other glacial forms are presented and explained with regard to their origin.

The presentation of the range of forms is systematic, depending on the forces and processes that shape a landscape. Thus, this chapter is concerned with the interrelationships that shape what is known as the glacial form treasure.

The analysis of a wide variety of surface forms, the examination and evaluation of geologically significant **outcrops**, and the linking of the findings obtained help to

W. Fraedrich (✉)
Hamburg, Hamburg, Hamburg, Germany
e-mail: wolfgang.fraedrich@hamburg.de

give an impression of the history of the development of a landscape. "Learning to understand landscape" is a goal for which this book should be a helpful guide.

▶ **Outcrop** The geologist defines an outcrop as a section of the bedrock. The rock becomes visible, the observer gets a first insight into the "geological substructure" of a landscape, recognizes structures, finds rocks, can "grasp" the geology of the landscape, look at it, photograph it, draw it, and so on. One will be able to formulate hypotheses about the possible formation, because only the additional information about the geological structure of a landscape brings clarity.

Outcrops can be found in many places in Central Europe. In the formerly glaciated or ice-marginal regions there are numerous sand, gravel, crushed stone and clay pits. There are often partly unvegetated road cuts, and on the sea coasts – especially along the Baltic coast – there are kilometres of cliffs. Especially there, you should go in search of tracks – although caution is advised, because the loose sediments or those soaked by the water can break off and slide down the slope! If gravel pits are still being mined, an access permit must be obtained from the operator beforehand.

The presentation of the wealth of forms is systematic, depending on the forces and processes that shape a landscape. The interrelationships that have already been addressed in Chap. 2 form the basis for the understanding of the so-called glacial form treasure.

5.1 Overview of Forces and Processes

In general, various forces and processes are involved in the formation of relief. In addition to all **endogenous** influences, **exogenous** forces and processes also play a role.

▶ **Endogenous Dynamics** Endogenous dynamics includes all forces and processes that are controlled from within the Earth. A very slow process is, for example, the formation of mountains, while earthquakes or volcanic eruptions are "geological momentary events". The "motor" of endogenous dynamics is heat, which is released, among other things, by the decay of radioactive material in the Earth's interior and released to the outside.

▶ **Exogenous Dynamics** The exogenous dynamics (their forces and processes) are controlled from outside. The control mechanisms result from the dynamics in the terrestrial air envelope, which in turn is a result of the different radiation budget in the individual latitudes. The "motor" of the exogenous dynamics is solar energy, which is converted into heat via the Earth's surface. The result is a zonally very different climate, which can also have relief-related regional characteristics.

Thus, varying temperatures and precipitation, but also varying air pressure conditions and, as a consequence, the winds in the lowest shell of the atmosphere (the troposphere with an average height of about 11 km) influence the type, extent and intensity of **weathering, erosion, transport** and **deposition**. These four processes take place in a perpetual cycle, and have done so for several billion years, precisely since the Earth has had an atmosphere. In this context, one speaks of the **morphodynamic cycle** (Greek *morphos* = the shape; *dynamic* = movement).

The preconditions for how quickly rock on the surface can weather are very different. On the one hand, the type of rock with its specific mineralogical and chemical properties plays a role, and on the other hand, weathering is dependent on the climatic parameters of precipitation and temperature.

In Chap. 4, the formation of a sinkhole was mentioned in connection with the Lieth lime pit. Here, solution weathering, i.e. a chemical process, has taken effect. Chemical weathering processes take place much more intensively at higher temperatures and with higher precipitation, and their effectiveness is of course also dependent on the chemical composition of the rock. However, the diversity of chemical weathering processes will not be discussed in depth in this book because they are largely insignificant in subpolar and polar climatic regions, but the background of the described sinkhole formation will be briefly explained.

The starting point of this process was soluble rocks, i.e. **carbonate rocks** such as limestone ($CaCO_3$) or dolomite ($CaMg(CO_3)_2$), **sulphate rocks** such as gypsum ($CaSO_4 \cdot nH_2O$) and **salt rocks** such as rock salt ($NaCl$). The intensity of the dissolution processes depends on the chemical purity of the rocks, the prevailing temperature conditions and the size of the outer and inner (cavities) rock surface that can be reached by the dissolving water. The rocks react very differently. While 1 cm of rock surface in limestone is chemically weathered in about 1000 years in a humid temperate climate, as is currently the case in our latitudes, it only takes 100 years for gypsum. Salt rocks are the most soluble.

The process of **carbonic acid weathering** plays a dominant role. Although carbonic acid is only very weak, it becomes particularly effective in highly soluble carbonate rocks. Chemically, the following process takes place: $CaCO_3 + H_2O + CO_2 \rightarrow Ca(HCO_3)_2$. Carbonic acid ($H_2CO_3$) is a chemical compound of the compounds water (H_2O) and carbon dioxide (CO_2) present in the atmosphere.

In addition, **hydrolysis** (solution weathering of silicate rocks such as granites or basalts) and **oxidation** play a role (reaction of the rocks with the oxygen in the air and the oxygen dissolved in precipitation and physically), in the course of which the rock undergoes a reduction in strength from the surface; the consequences are **corrosion** and discoloration (as in the case of rust formation, for example, up to soil formation or the formation of unconsolidated rocks, in the tropics this leads to the formation of lateritic soils).

Much more important in regions characterized by ice climates are mechanical (= physical) weathering processes. And among these, the process of **frost heave** plays the most important role. The determining factor is a repeated temperature change around the freezing point (0 °C), which is typical above all for subarctic climatic

Fig. 5.1 Frost blasting: Block debris in the high mountains in Upper Carinthia/Austria (© Wolfgang Fraedrich)

regions and for areas outside the ice. The water penetrating into fissures and cavities of the solid rock freezes, thaws again, freezes again, etc.. In the process, the volume of the water changes again and again. At +4 °C it has the smallest volume, during crystallisation to ice the volume can increase by up to 9% (at −25 °C). This exerts a considerable explosive pressure on the surrounding rock surfaces (e.g. at −25 °C about 2400 kp/cm^2).

The weathering products depend on the rock, e.g. sandstone results in sand, while granites, gneisses or basalts initially weather to coarser rubble (Fig. 5.1). The end product of frost heave is the coarse clay (equivalent diameter 0.0006–0.002 mm according to DIN 18196, the soil classification for civil engineering purposes).

The effective range of frost heave depends on the depth of the frost heave. In most cases this weathering process is limited to depths of up to 2 m, in exceptional cases in permafrost areas (cf. Chap. 7) also up to 200 m.

Frost blasting also occurs at the base of a glacier. This makes it easier for flowing glacial ice to pick up "blasted" rock fragments at its base as well as on the sides of a valley glacier. In addition, wherever glaciers are surrounded by rocky mountains, weathering debris falls onto the ice surface. This initially flows with the ice on the surface, is then covered by fresh snow and is passively included in the ice formation process. Since the glacier loses mass at its base due to the flow process, but gains mass at the surface above the snow line due to fresh snow, the debris travels deeper and deeper along its path. Even when it reaches the base of the glacier, it is then transported further by the ice and thus usually receives a slight rounding by grinding on the bedrock.

After this brief introduction to fundamental contexts, the specific shaping processes as well as the resulting set of forms will be presented in more detail in the following.

5.2   The Glacial Mould Record

Flowing ice is the morphologically effective force (Fig. 5.2), and the processes triggered by this force are ablation, transport and deposition. The ablation processes are in turn subdivided into **detersion, detraction** (Fig. 5.3) and **exaration**.

▶ **Detersion** Detersion (lat. *deterere* = to grind) refers to the stress on the solid rock bedrock caused by the solids of any grain size carried along at the base of the glacier.

▶ **Detraction** Detraction (lat. *detrahere* = to pull out) is the breaking out of the frozen rock parts at the base of the glacier.

▶ **Exaration** Exaration (lat. exarare = to *rut*) is defined as the excavation, scaling and folding of preexistent loose and rocky material at and in front of the glacier front.

The result of these processes are numerous relief forms, which can be assigned to two different form groups, the glacial ablation forms and the glacial deposition forms.

5.2.1   Glacial Erosion Forms

Two typical forms are found in mountain areas, the glacial **cirque** and the **trough valley**. The cirques (Fig. 5.4) were the origin of mountain glaciation. Here, snow initially accumulated in a trough, which over the years became firn and glacial ice, then soon filled the trough and finally began to flow downhill. Detersion and detraction were thus able to set in, the subsoil deepened and the slopes steepened until finally a "armchair shaped" hollow form was created.

In the further course, the ice has flowed through already existing valleys formed by rivers. In the process, rock was removed both at the base (**deep erosion**) and at the sides (**lateral erosion**). Detersion and detraction also played a decisive role here. The result of an erosion process that lasted for centuries or even millennia is the transformation of the originally V-shaped **valley** cross-section into a U-shaped trough valley (Fig. 5.5).

Such trough valleys are found in all high mountains of the earth, thus also in the Alps, but also in all Pleistocene and recent glaciated mountain areas of Scandinavia,

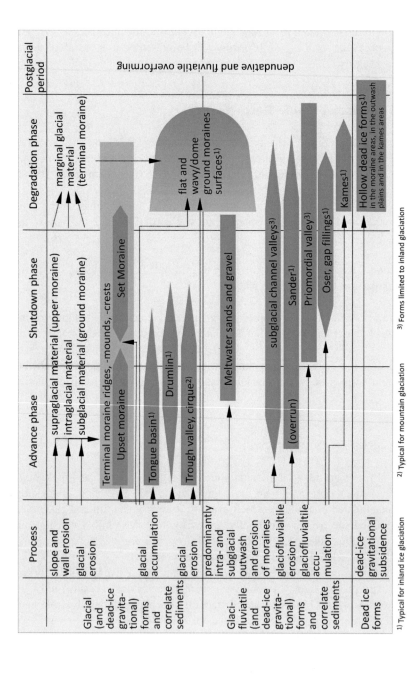

Fig. 5.2 An overview of the glacial mould record (© Wolfgang Fraedrich)

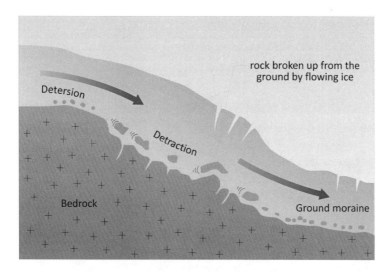

Fig. 5.3 Detersion and detraction (© Wolfgang Fraedrich)

Fig. 5.4 View into a glacial cirque in the Hohe Tauern: In the now ice-free hollow form there is a cirque lake at the deepest point (© Wolfgang Fraedrich)

Fig. 5.5 A trough valley in the north of Iceland near Ólafsfjörður (© Wolfgang Fraedrich)

especially in Norway including Spitsbergen, on Greenland and on Iceland, as well as of course at the edge of the continent of Antarctica.

On the long way to the lower lying regions, the bedrock has often been literally "planed". In large parts of southern Scandinavia, for example, you will find rounded, often even scuffed rock **humps**, so-called round humps or round humps. If you take the ferry from Denmark or Germany to Larvik (Norway) or Stockholm (Sweden), for example, you will pass a large number of small, often naked rocky islands on the last section of the ship's passage.

Many of them show the typical detersive glacial scars, the course of which at the same time indicates the direction of flow of the glacial ice at that time. The fact that many of these humps lie in the sea today can be explained by the late and post-Pleistocene eustatic sea-level rise (cf. Chap. 8).

When the ice reaches shallower areas, its flow velocity decreases due to the lower gradient, and a transition from detersive to exarative scouring occurs in the marginal zones of the ice sheet.

With an adequate supply of ice from the nutrient zone, the marginal zones can have ice thicknesses of several hundred metres. Here, too, the ice flowed in pre-existent shallow hollow forms and reshaped them into so-called **tongue basins**. Such tongue basins are typical for the Pleistocene glaciated areas of Central Europe. In Schleswig-Holstein as well as in Mecklenburg-Vorpommern and Brandenburg, but also in Bavaria and Baden-Württemberg, there are numerous of these ablation forms, which are especially pronounced in the area of the former ice margin and are arranged there next to each other. Today, the deeper areas of these tongue basins are often filled with lakes (e.g. Plöner See, Chiemsee, Starnberger See or Bodensee).

5.2.2 Glacial Depositional Forms

On the long way to the outermost ice margin, which was more than 1500 km long for the Nordic ice sheet during the Elster and Saale Glacial Periods and more than 1000 km long during the Weichselian Glacial Period, enormous amounts of rock material were picked up, transported and deposited again. Glacial geologists refer to the glacial deposits as moraine (Fig. 5.6). It consists mainly of clay and coarser **gravels**.

▶ **Clay** Clay is the term used to describe a substrate composed of individual particles of varying grain size. These include clay (<0.002 mm), silt (0.002–0.063 mm) and sand (0.063–2.0 mm). The loams sedimented by the inland ice in the North German Lowlands are also called boulder clay (if calcareous).

▶ **Debris** Bedloads are rock fragments of various sizes, some of which can weigh several tons, transported by glacial ice and mechanically stressed by the sliding of the ice along the rocky bedrock. Large boulders are very numerous in the north German area, they are called erratic blocks (Fig. 5.7). Since the rock type of the boulder is identifiable, its place of origin can be clearly reconstructed in many cases. Such boulders are called "Guide bedload". The bedload analysis of Pleistocene ground moraines thus allows a reconstruction of the glacial movement at that time.

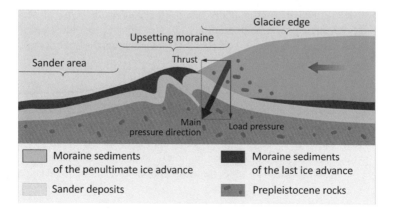

Fig. 5.6 Longitudinal section through the end of a glacier (© Wolfgang Fraedrich)

Fig. 5.7 The "Wandhoff-Findling" in the erratic boulder garden near Kreuzfeld (near Malente-Gremsmühlen) (© Wolfgang Fraedrich)
The 4.00 m high, 4.20 m long and 3.50 m wide giant erratic boulder weighing about 126 t was discovered in 1983 in the nearby gravel pit of the Wandhoff company (hence the name "Wandhoff erratic boulder") and was moved to its present position in 1990. This Småland granite is located in southern Sweden, it has an age of about 2 billion years

5.3   Reconstruction of Glacier Movements by Guiding Bedloads

Bedload research is a special branch of Quaternary geology. Its aim was and is the determination of boulders and the reconstruction of their origin. Once the bedloads have been determined petrographically, i.e. with regard to the type of rock, their origin, their percentage composition in the moraine deposits, and the **emplacement conditions have** been established, it is possible to reconstruct the direction of flow of the inland ice of that time.

▶ **Control Measurement** The position of the longitudinal axis of the bed load in the sediments is determined by means of situmetry (cf. Sect. 5.4). This requires the use of a control table (or situmeter). With the help of a compass the position in space can be determined concretely. An average value of the long axis alignment corresponds to the flow direction of the ice, because bedloads align themselves during the flow process in such a way that they have the lowest possible frictional resistance.

Such investigations have been carried out in large numbers in the Pleistocene glaciated areas (Fig. 5.8). Since the ice streams of the different cold periods were interrupted by warm periods and thus correspond to independent and also different

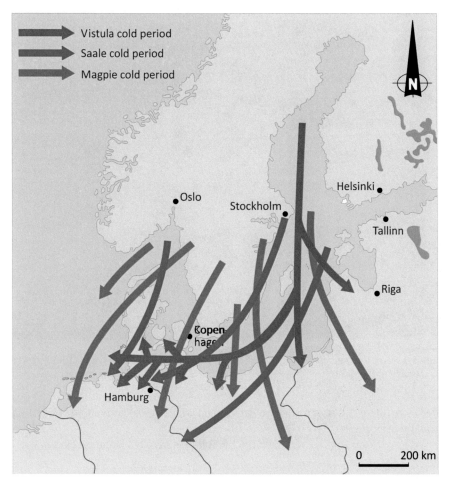

Fig. 5.8 Reconstruction of different ice advance directions during the last three cold periods in north-central Europe on the basis of lead slates (© Wolfgang Fraedrich)

ice streams, their bedload content has a specific composition in each case. The same applies even to the different ice advances within a cold period.

5.4 Simple Investigations on a Moraine Outcrop

If one arrives at an outcrop in a Pleistocene glaciated area, for example in north-central Europe, there are various ways of reconstructing the origin of the landscape section. One of them is an analysis of the outcrop structures and the composition of the rock (Fig. 5.9).

Since the sediments could – theoretically – have been sedimented by four different forces, namely by the ice, by flowing water, by ocean currents in the coastal

Fig. 5.9 Moraine outcrop on the south coast of the island of Rügen near Klein Zicker
(© Wolfgang Fraedrich)

area or by the wind, it must first be determined which so-called **transport agent** is
actually responsible.

When looking at a moraine outcrop, the following typical features stand out:

- The sediment body shows no stratification. The sediments are predominantly
  clay, which can be both strongly sandy and very fine-grained.
- Interbedded are numerous petrographically varying stones of different sizes,
  which are predominantly edge-rounded.
- The sediment body appears relatively compact in itself.
- In the coastal area (e.g. along the cliffs on the Baltic Sea coast), many rocks of
  different types and sizes are also found at the foot of the cliff, some of which are
  already (an)rounded due to the rolling movements in the surf area.

Due to the presence of the coarse, edge-rounded debris and the lack of stratification,
deposition by flowing water and wind can already be ruled out. In addition, marine
deposition, i.e. deposition caused by ocean currents, can also be ruled out, since
coarse bedloads cannot be moved by these either. The knowledge gained in this way
could now easily be confirmed with a geological map (as boulder clay or marl).

If information on the direction of flow is required, bedload alignment measurements
(Figs. 5.10 and 5.11) are helpful. The longitudinal axis alignment of at least
100 bedloads should form the basis of such a measurement process.

Fig. 5.10 Adjustment measurement in the field (© Wolfgang Fraedrich)

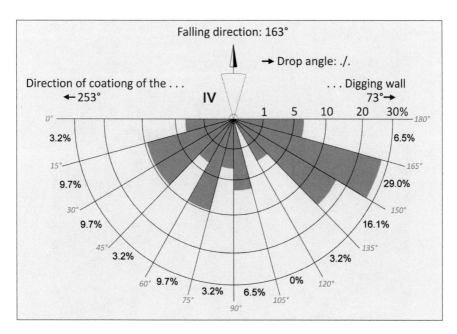

Fig. 5.11 Example of the graphical representation of a closed-loop control measurement (© Wolfgang Fraedrich)

5.5   Moraines in North-Central Europe

In the landscape, moraines show up in very different ways. For the area of north-central Europe and parts of southern Scandinavia (especially eastern Denmark, the Danish Baltic Sea islands, but also for Scania, the southernmost region of Sweden) we distinguish three Pleistocene formed moraine types (Fig. 5.12) which are

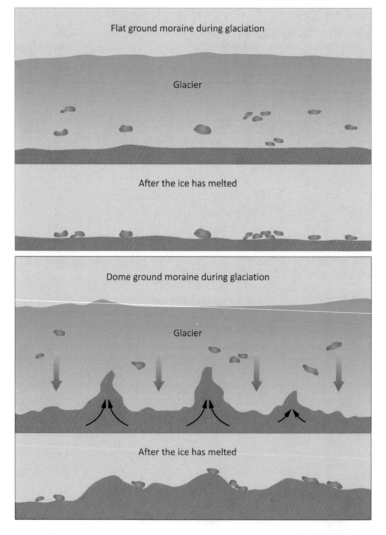

Fig. 5.12 Flat and dome-shaped ground moraines in development (© Wolfgang Fraedrich)

- shallow to undulating ground moraine,
- the dome-shaped ground moraine,
- the wall-shaped terminal moraine.

All three moraine types dominate the relief in the Weichselian regions, the so-called **young moraine landscape**. In the regions of central Schleswig-Holstein, Lower Saxony, Saxony-Anhalt and southern Brandenburg, the so-called **old moraine landscape**, these types of form have already been deformed by weathering, erosion and redeposition processes during the Eemian and Weichselian periods due to their greater age of over 120,000 years. Only the main ice margin of the youngest main ice advance of the Saale glacial period, that of the Warta stage, is still pronounced as a prominent ridge.

Figure 5.13 shows a map section of East Holstein, which shows the three landscape types mentioned in their typical sequence. In the northeast, for example on the island of Fehmarn and in the region around Oldenburg/Holstein, the flat to undulating ground moraine landscape predominates. Anyone who has ever been to the island of Fehmarn will remember that this is an island without any elevations worth mentioning, and further to the northeast, on the Danish islands of Lolland and Falster, one finds a comparable relief. Further to the southwest, there is a transition to the dome-shaped ground moraine landscape, which is finally delimited by a conspicuous range of hills, the terminal moraine.

The difference between these individual moraine types is due to different exarative processes. The flat to undulating ground moraine landscape developed under the ice (= subglacial) in a zone far away from the former ice margin. The thickness of the ice was in parts many hundreds of metres, the pressure of the ice on the bedrock was great, and the ice had hardly any crevasses. The material carried and pulverized over long distances by the inland ice remained here after melting. Towards the ice margin, the thickness became smaller, the ice pressure on the bedrock was clearly lower, and crevasses often formed due to the stresses in the ice body, which was only plastic to a limited extent, especially as smaller unevennesses in the terrain had to be overcome. If the inland ice flowed through trough-shaped landscapes, expansion crevasses formed at the base of the glacier, which were often widened by meltwater. Even through the now clearly lower ice pressure, the moraine debris that had accumulated on the surface at the base of the glacier was still pressed into the crevasses from below, so that after the complete melting back, the hilly to dome-shaped landscape remained.

While the ground moraines are deposits and relief forms of subglacial origin, the end moraines (Fig. 5.14) were formed at the immediate ice margin. Terminal moraines are erosional debris that has been reworked by detritus and exaration and transported in front of the glacier margin.

At the front of the glacier, distinctive ramparts were formed, which rose above the surrounding area after the ice melted back. Depending on the mass balance of the inland ice, a distinction is made between the **set end moraine** and the **upsetting end moraine**. During the formation of the set end moraine there was a balanced mass balance, i.e. a balance between the melting of the ice in the feeding area and the ice

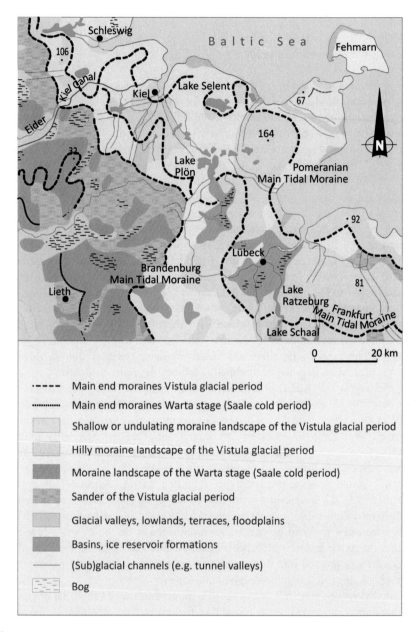

Fig. 5.13 Landscape in Ostholstein (© Wolfgang Fraedrich)

replenishment in the nourishing area, so that the material carried along with the flow was merely deposited at the ice margin. As a result, set end moraines only reach low heights and hardly rise more than 30–40 m above the surrounding area. The accumulation of compressive moraines always occurred when the ice was still

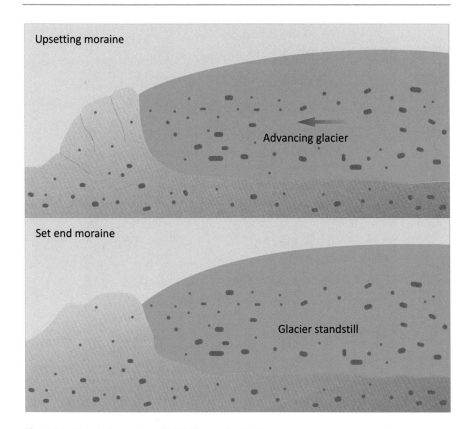

Fig. 5.14   Terminal moraines (© Wolfgang Fraedrich)

advancing, i.e. when the mass balance was positive. More and more material was thrown up in front of the glacier front and compressed by the continuing sliding pressure of the ice. The basic principle of this process can be compared to a bulldozer (= inland ice) pushing soil (= moraine material) in front of it.

5.6   Moraines in the Mountains and Mountain Margins

The Pleistocene glaciated areas of the Alpine fringe show similarities with the moraine types in north-central Europe, but the shallow ground moraine is not represented here due to the smaller extent of the Alpine glaciation and the much more pronounced relief of the subsoil. The dome-shaped ground moraines form the direct transition from the tongue basin lake to the terminal moraine walls, which enclose the tongue basins almost in a semicircle.

In some cases, individual tongue basins, such as that of the Inn Glacier and that of the Chiemsee Glacier, border directly on each other. Here the Alpine glaciers flowed out into the foreland via various valley openings (Inntal and Achental) and then

Fig. 5.15 Lateral glaciers at the edge of Heillstugubreen in southern Norway with pronounced lateral moraines marking the glacier level in the mid-nineteenth century (© Wolfgang Fraedrich)

reunited to form a larger system immediately after leaving the mountains. As a result, the entrained moraine material was merged into a **medial moraine** at the margin of two glacier tongues. This can be seen in an impressive way on a drive along the A 8 federal motorway (Munich – Salzburg): Coming from Munich, one very soon leaves the Munich gravel plain and crosses the western terminal moraine of the Inn glacier area at Irschenberg. In the further course one crosses its tongue basin. At the Frasdorf junction, you cross the medial moraine of the two glacier areas and enter the Chiemsee glacier area, which you finally leave again in the area of the Traunstein junction when you cross the eastern terminal moraine.

In the recently glaciated areas, e.g. of the Alps, Scandinavia (Iceland, Greenland, Norway) and North America (Alaska), the narrow and elongated valley glaciers often build up elongated **lateral moraines** (Fig. 5.15). Due to the interim ice peak in the mid-nineteenth century and the subsequent melting of the glaciers, the lateral moraines marking the maximum ice advance of that time are now visible as conspicuous ridges in the landscape.

5.7 Other Glacial Forms

Both in the marginal area of the Pleistocene Nordic Ice Sheet and in the area of the former Alpine foreland glaciers, two further glacial forms appear, **drumlins** and **dead ice hollows**. Drumlins (derived from the Irish word *druin* = small hill) are elongated, "streamlined" ridges consisting of ground moraine material. Two causes are now thought to be responsible for their formation. Firstly, the ice may be overloaded with moraine material towards the ice margin, which is then deposited at the base and drawn out by the ice flowing onwards, thus reshaping it. Secondly, in the case of a renewed ice advance, an already existing dome-shaped ground moraine can be overformed accordingly. Drumlins rarely rise more than 10 m above the surrounding area after the ice has melted back, they are up to 50 m wide and usually up to 200 m long. Typical for this subglacial depositional form is a steeper windward side and a more gently sloping lee side. In many cases, drumlins occur in clusters and are referred to as drumlin fields, e.g. near Bad Oldesloe in Schleswig Holstein or southeast of Lake Starnberg, where more than 100 drumlins form the Eberfinger drumlin field (Fig. 5.16).

Fig. 5.16 Drumlin in the Eberfinger drumlin field southeast of Starnberg (© Wolfgang Fraedrich)

The identification of such drumlins is already possible by looking at the outer shape. Dome-shaped ground moraines are not elongated, end moraines or medial moraines protrude higher. If several such small ridges occur parallel to each other, i.e. their longitudinal axis points in the same direction, the interpretation is clear. Final clarity is provided by the analysis of the geological structure, which becomes possible when a drumlin is exposed: the outcrop shows the structure and composition typical of a moraine.

Dead ice hollows also form in the depositional area of glaciated regions. When a glacier begins to gradually melt back, blocks of ice are often left behind at the edge of the ice because crevasses either from the glacier surface or from the glacier base completely "thaw". While even the gradually melting glacier still flows forward, the ice blocks left behind are no longer involved in this flow process, hence they are called dead ice blocks. In the **Late Glacial**, i.e. towards the end of the Weichselian, they often filled the lower lying areas of the dome-shaped ground moraine landscape, but melted only very slowly.

They were even covered by sediments supplied by meltwater from the ice margin and thus preserved from further thawing. It was not until the beginning of the Alleröd warm period around 10,500 years BC that the thawing of the dead ice began, which lasted until the Preboreal, i.e. until around 8000 BC. Due to the sealing of the hollow forms, the originally glacially created relief was largely preserved. In many cases, lakes had formed over the hollows filled by the dead ice during the subarctic period, and these lakes had frozen over during the long winter period, so that even the finest material was deposited. Today, this forms an impermeable layer, so that many of the dead ice hollows are filled with lakes, so-called dead ice lakes (Fig. 5.17), even in the present day. Some lakes, especially smaller ones, have already silted up or become boggy during the post-glacial period.

Particularly in the young moraine country of Schleswig-Holstein and Mecklenburg-Vorpommern, small hollow forms can be found in the middle of farmland, usually surrounded by a group of trees. These are also such dead ice holes, which are called **Soll** (pl. Sölle, from Plattdeutsch Soll = water hole) in technical language. In the course of land consolidation measures, especially in the 1960s, many of them were simply filled in, but with little success, because the impermeable layer leads to waterlogging, so that the agricultural use of these small areas brought no yield.

Fig. 5.17 Lake Totei east of the Hüttener Berge in central Schleswig-Holstein. In the foreground the terrace surface of the younger subarctic time. (Younger Dryas period, 12,000–11,500 years before present) (© Wolfgang Fraedrich)

The Influence of Glacial Meltwater

6

Wolfgang Fraedrich

Abstract

Comparable to the dynamics of glacial ice, there are also dynamics of (glacial) meltwater. Here, too, it is important to understand the effective forces themselves in order to be able to apply them to the understanding and reconstruction of so-called fluvioglacial processes.

Meltwater systems, like glacial ice systems, build up a typical set of forms; here, too, it is important to distinguish between ablation forms and depositional forms.

This chapter provides a wealth of information for understanding fluvioglacial processes and also illustrates that there are also form societies that have been shaped by both glacial and fluvioglacial processes.

Here

- the reader learns in what way flowing water has a relief-defining effect,
- the reader learns to distinguish between forms of erosion and forms of deposition and to understand their origin,
- the reader gets hints how to reconstruct fluvioglacial dynamics in gravel or gravel pits,
- the reader learns via the simplified model of the "glacial series" under which conditions such form societies develop,
- is used as an example to explain how a glacially and fluvioglacially shaped landscape can be described and interpreted on the basis of geological maps.

W. Fraedrich (✉)
Hamburg, Hamburg, Hamburg, Germany
e-mail: wolfgang.fraedrich@hamburg.de

It has been repeatedly pointed out that glaciers constantly release meltwater. The water released at the surface as a result of an increase in external temperature to above freezing point and at the base due to the flow movement of the glacier is also a formative element in the so-called morphodynamic cycle. It is able – according to the basic laws of fluviatile (= flowing water) dynamics – to erode and redeposit. These fluvioglacial (= glaciofluvial) processes take place both in and under the ice (= intra- and subglacial) and outside the ice.

6.1   Overview of Forces and Processes

In order to understand the basic relationships described below, the morphological effectiveness of flowing water should first be explained. Since meltwater behaves in the same way as flowing water in general, the laws derived for rivers in other climatic zones can be transferred. The dynamics of flowing water occur in all climatic zones. It only varies in its effectiveness due to the different quantities of water in total and their seasonal distribution. For the subpolar regions of the earth – and this is also known to be true for the Pleistocene ice margins – a short-term, periodically even very strong runoff of snow and glacial melt waters is decisive. The debris originally released by weathering processes has already been transported over long distances by the ice, and in the process has already been subjected to considerable mechanical stress.

The relief-shaping effect of flowing water is expressed in two ways. Firstly, by erosion both vertically (deep erosion) and laterally (lateral erosion) and secondly by deposition. Erosion takes place through the lifting and transport of loose rock particles and through corradiation, i.e. through the abrasion of the rock in the riverbed by the solids carried along. Fluviatile relief formation is determined by various components:

- the drag force of the flowing water, whose energy can be approximately determined with the formula: $E = 1/2\ mv^2$ (m = mass of water, v = flow velocity),
- the discharge of the watercourse,
- the quantity and size of solids carried, and
- the resistance of the rock in the erosion zone of the stream.

Figure 6.1 clearly shows how the morphological effectiveness of flowing water develops as the waterway increases. Transferred to the area of fluvioglacial processes, this means that deep erosion takes place especially subglacially, while after the meltwaters have left the glaciated area, lateral erosion gradually predominates and is soon replaced by sedimentation.

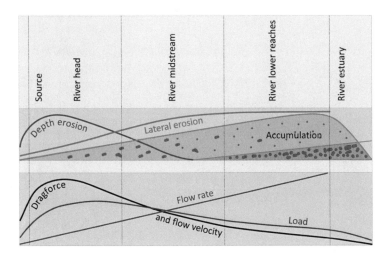

Fig. 6.1 Behaviour and effects of flowing water–simplified (© Wolfgang Fraedrich)

6.2   Fluvioglacial Forms of Erosion

Under a glacier, and here especially under large inland ice glaciations, entire power grids form. In some cases, they also erode ice at the base of the glacier, so that regular tunnels or even tunnel systems can develop in the process. Depending on the flow velocity and the bedrock, the glacial relief is thus fluvioglacially reshaped. So-called **subglacial gullies** or **gully systems of** very different dimensions are formed. Figure 6.2 shows a landscape section in western Mecklenburg-Vorpommern between the Baltic Sea and the Elbe valley. The generalized representation of the glacial-geological structures shows a gully system running almost in north-south direction, in the middle of which lies the Schwerin lake district and which runs according to the main direction of advance of the Weichselian ice sheet in this region. Its continuation to the south runs outside the Weichselian glaciated area in a widening valley. It is certain that this landscape, as can be seen from the main ice margins, was decisively shaped by glacial action, and the subglacial deformation by meltwater gave the final "finishing touch". The Schwerin lakes also have the character of a tongue basin, but in their present form, also with respect to the water depth, they can rather be described as a subglacially formed channel lake system. The same applies to the Schaalsee, for example, which is embedded in an almost parallel channel system.

   The glacial geologist calls the valley forms created under the ice **channel valleys**, and the lakes embedded in them **channel lakes**, which are narrow and elongated in their outer shape, sometimes also have a very irregular course of the shore, and which also have the greatest water depth according to the former **course of the stream**.

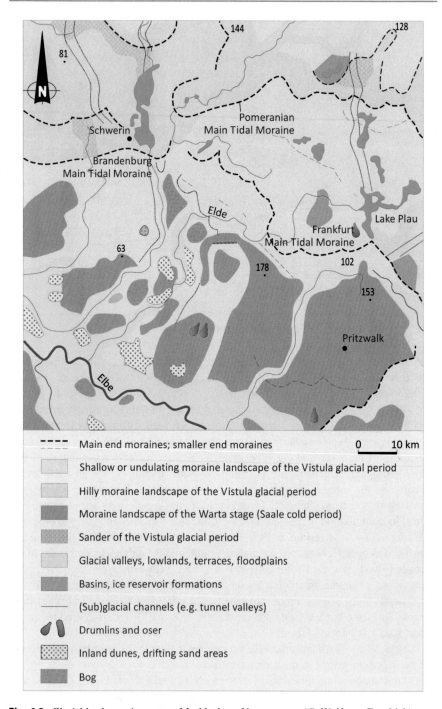

Fig. 6.2 Glacial landscape in western Mecklenburg-Vorpommern (© Wolfgang Fraedrich)

▶ **Streamline** The streamline is the area of a watercourse where the greatest flow velocity occurs. In a symmetrical river cross-section, this is greatest in the middle of the river. Irregularities in the river bed lead to a deflection of the line of greatest flow velocity to one of the two sides (where there is less resistance) and, as a result of the erosive forces (depth and lateral erosion are the consequence), to an ever greater asymmetry of the river cross-section.

Outside the glaciated area, the subglacial channels merge via meltwater valleys into the **receiving waters,** the valley into which various meltwater streams flow. Corresponding to the different main ice margins during the Saale and Vistula glaciations, which mark standstill phases over a longer period of time, various such receiving waters – in technical terminology they are called glacial valleys – can be distinguished (Fig. 6.3). Over many millennia, millions of cubic metres of meltwater have combined in them to form large streams that created wide valleys. During the standstill of the glaciers and during the gradual melting back, these streams formed enormous primeval current systems in these valleys, in which – depending on the water flow and thus even within a year or even a few months – new river branches developed again and again, while others were "shut down" by deposits. By no means correct is the idea that the entire valley was the riverbed at that time, i.e. completely filled with water.

On the way to these glacial valleys, the meltwater collected in lower-lying and initially drainage-free cavities. At the beginning of the melting process, these were usually located within the formerly glaciated area between the ice margin and the terminal moraine wall. Figure 5.1, for example, shows the relict of such a former ice dam landscape southwest of Lübeck. The hollow form as such is not of fluvioglacial origin, but was formed by the exaration of the glacier. Due to the long winters in the subarctic region at that time, the lake surface could freeze over, so that the finest sediments carried by the meltwater could settle in the water, which remained still for several months. Therefore, clays form the basis of such former ice reservoirs.

6.3   Fluvioglacial Depositional Forms

Starting from the subglacial area, the oser (sing. **Os**) should be mentioned first. Like the drumlins, these are narrow, elongated, wall-like ridges. In contrast to the drumlins, however, they are arranged in rows and are often much longer, sometimes up to three kilometres. Their formation is due to the deposition of sands and gravels in intra- and subglacial meltwater channels, whereby the intraglacial deposits were gradually deposited on the ground moraine material as the ice thawed. According to the prevailing flow direction of the meltwater at that time, they do not have to run in a straight line, but can also show slight bends.

In marked contrast to the drumlins, the inner structure of an Os does not show any moraine material, i.e. no completely unsorted boulder clay or boulder clay with embedded coarser boulders, but predominantly well-rounded sands and gravels, which are also recognisably stratified, as is usual with fluviatile sediments.

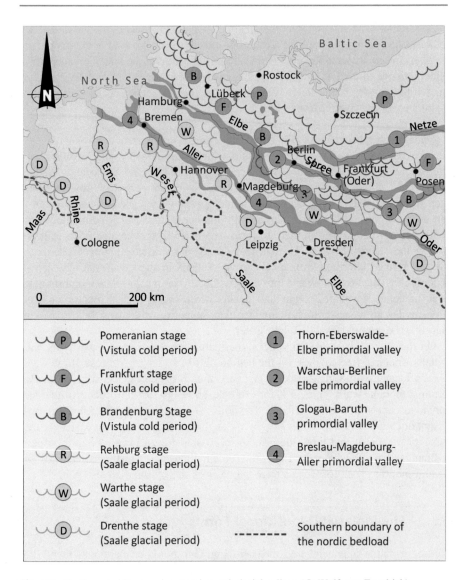

Fig. 6.3 North-central Europe: ice margins and glacial valleys (© Wolfgang Fraedrich)

In the former marginal area of the ice, where the meltwater flowed through many crevasses, **kames**, deposits of sand and gravel, formed in many places.

Kames are not always easy to spot in the landscape. They are usually small roundish hills only a few metres high, arranged in groups at random and separated from each other by valley or bowl-shaped depressions. They are smaller in extent than dome-shaped ground moraines. If one has – e.g. by orientation on a geological map – indications of the former ice margin and finds such forms in such a region, it is obvious to call them kames. In addition, such kames can usually be recognized by their vegetation, because the sandy, thus nutrient-poor and water-permeable soils often allow only sparse grass growth or a stand of predominantly coniferous trees.

Final clarity is also given here only by an outcrop. Then one recognizes a stratified structure of predominantly rounded sand and gravel, whereby the gravels – differently with the Osern – are not too coarse, since the fluvioglacial system was very small-scale structured and the flow velocity was therefore not always very large.

The largest area of fluvioglacial forms is occupied by the **sands** (Icelandic: sandur). Wherever the meltwater streams preceded the ice margin – and this was always the case along the main ice margins – they usually branched off into a multitude of individual partial streams until they finally flowed into the glacial valley. Since the water flow varied greatly due to the seasonal climatic changes, erosion and deposition alternated in a very small area. Over the years, sometimes several decimetres thick sediment bodies were built up, which were relatively flat in themselves, within which the rivers shifted again and again. Figure 6.2 shows a huge outwash plain south of the Weichselian ice margin near Schwerin; in terms of area, it is one of the largest in northern Germany. In southern Germany the term "Sander" is not commonly used, but rather **"Schotterebenen" (gravel plains)**. This conceptual differentiation results from the composition of the deposits. Whereas in the north-central European region, sands and gravels predominantly build up such outwash plains, in the foothills of the former Alpine foreland glacier areas they are much coarser gravels and pebbles due to the much shorter transport distance. Probably the most extensive contiguous gravel plain extends between the Isar-Loisach glacier area and the Inn-Chiemsee glacier area: the Munich gravel plain, an almost completely flat landscape that has only been linearly deepened by the Isar.

Even if the interpretation of an outwash plain or gravel plain is relatively clear from the relief alone, an insight into its geological structure offers much of interest. Some suggestions are given in the following.

6.4 Analytical Methods for the Examination of Gravel Pit Outcrops

Both in northern Germany and in the Alpine foothills, sand and gravel pits or gravel pits are very numerous. The material extracted there is almost inexhaustible and represents an important basis for the construction industry (including road building, concrete production, etc.). Only the strong and often widely visible interference in the landscape keeps people from really mining on a large scale. The condition of the pits varies greatly. While many smaller ones have already been abandoned, i.e. are no longer used and therefore grow over again and form their own ecotopes, there are a large number of pits of larger dimensions in which mining has taken place right down to the groundwater horizon and which have "drowned" after mining operations have ceased. In many places, the so-called "dredging lake" prevents a view of the structures created by nature. More interesting are all those pits that are accessible and show freshly cut pit walls (be careful with steep, high walls, there is

Fig. 6.4 Sander deposits near Norderstedt, Schleswig-Holstein (© Wolfgang Fraedrich)

danger of falling down!). In principle, an access permit must be obtained from the operator.

Substrate analysis of sanders and gravel plains reveals much about the geological development of the time (Figs. 6.4 and 6.5). If the pit lies at the root of a sandy or gravel plain, where meltwater originally emerged from the glacial zone, the coarsest sediments are to be found. If the gravels, which are readily visible in external form, have a relatively high proportion of edge-rounded components, this is evidence of deposition near the terminal moraine, for the transport route was not yet long enough to permit complete rounding of the particles. In pits with increasing distance from the former ice margin, the sediments in place become smaller and smaller in composition, and the degree of rounding increases more and more with increasing distance from the terminal moraine due to the longer transport path.

In addition to the size and degree of rounding of the sediments, the course of the individual layers is of interest. If the layers lie almost parallel to each other over

Fig. 6.5 Gravel deposits near Huglfing, Isar-Loisach glacier area (© Wolfgang Fraedrich)

greater thicknesses (the geologist then speaks of concordance or concordant bedding), this is an indication that the flow direction of the meltwater stream has changed little or not at all over a longer period of time. If, however, the strata run "in confusion", i.e. if they dip in different directions (the geologist then speaks of unconformity or discordant bedding), it can be deduced from this that in relatively short intervals of time the rivers and streams flowing in the outwash or gravel area have shifted their river beds and thus adopted a different direction of flow and thus of sediment deposition. On only a few square metres of pit wall, constant river bed changes can be traced in many gravel pits.

The thickness of individual sediment layers allows conclusions to be drawn about the varying flow velocities of the meltwater. The rule is that finer-grained layers alternate with coarser-grained layers. For the most part, this represents the seasonally varying amounts of debris. While in the winter months considerably less meltwater flows and therefore only the finer particles can be moved due to the lower transport force of the water, during the summer phase the coarser particles are also moved due to the considerably larger water volume.

The exploration of all these phenomena is possible without much technical effort. All that is needed – apart from permission to enter the mine site and clothing suitable for working on the terrain, including solid, sturdy footwear – is a spade. This enables you to uncover the structures. If you want to be even more precise, you can, for example, use a knife to expose the tops of individual layers where the strength of the mine wall permits, and use a (geologist's) compass to determine the direction of the steepest slope and thus the direction of the fill.

The situmetry method (Sects. 5.3 and 5.4) can also be used for meltwater deposits. However, the results are generally more difficult to interpret because a clear settlement is rather untypical.

6.5   The Composition of Sediments Tells Us Something About Their Origin

A closer look at gravel and crushed stone pits reveals a variety of different rocks. In particular, the deposits in the gravel pits of the Alpine foothills make it possible to develop more precise ideas about the origin of the rocks, because here the distances to the original rock are significantly smaller than in the gravel pits of north-central Europe. A regional classification is also easier because the gravel deposits in the Alpine foothills – unlike the sands and gravels in northern Germany – are bound to clearly delimited glaciers. Glacial debris and fluvioglacial sediments of a glacial system belong together, they are correlates (from correlate = to belong together), as the geologist says. If one has knowledge of the rock structure of, for example, the central Alps and the northern edge of the Alps – a geological map is often helpful here – then one is in a position to define the catchment area of the glacial system. It is not surprising, for example, that the gravel pit near Huglfing contains a relatively high proportion of Wetterstein limestone, which forms the Wetterstein Mountains with the Zugspitze massif. Ice and meltwater have removed large amounts of rock here and transported it to the region around Huglfing.

6.6   Form Societies of Glacial and Fluvioglacial Origin

Whoever walks or drives through landscapes shaped by the cold ages encounters a great impressive variety of forms. Rarely does the relief show a uniform picture, the surface forms change in the smallest of spaces. A ridge like that of a terminal moraine, over which a road leads, can be overcome in a short time even by bicycle, and when driving through a dome-shaped ground moraine landscape, the picture changes constantly. It goes up and down again and again, winding paths and roads adapt to the relief. Often numerous lakes also characterise the landscape. Whoever wants to think about the development of the landscape will at the same time have to "mentally sort" different observations in order to let them become the picture of an understandable complex whole.

6.6.1   Suggestions and Aids for Landscape Interpretation

The interpretation of landscapes requires practice. It does not succeed the very first time. But if you keep trying to walk through the landscape "with your eyes open", you will soon find that your knowledge and experience develop. Regardless of this, there are a number of ways to approach the understanding of a landscape in a targeted manner:

- Record and attempt to describe the forms in terms of their external shape (extent, height, position in the overall relief).
- Look for evidence of the geological subsurface.

If no outcrops can be found, the vegetation often already gives clues to the nature of the subsoil, because very fine-grained, clay-rich and thus more nutrient-rich soils develop on boulder clay or boulder loam than on the coarser-grained, clay-poorer soils of the outwash plains or gravel plains. Most of the terminal moraines are covered with mixed forests (in the young moraine country of north-central Europe the lime-loving beeches dominate), the dome-shaped ground moraine landscapes are either used as forests (mixed forests) or for pasture farming, while flat ground moraine landscapes are preferred areas for arable farming. Sand and gravel areas are used in different ways. In addition to forests (predominantly coniferous forests) and pasture farming, sandy soil-loving crops (e.g. rye, potatoes) are also cultivated.

- Work with cards.

If you are preparing more intensively for an exploratory walk, it is best to obtain appropriate topographic maps (scale at least 1:50,000, better still 1:25,000), because these show where road cuts and pits are to be found by means of an appropriate signature. In the digital age, for example, it is very easy to find out about https://www.openstreetmap.org in the field online.

The classification of a landscape in the regional geological context, which goes beyond topographic orientation, is possible by means of a geological map. It usually has the disadvantage of being structured in a very differentiated way, "peppered" with technical terms and therefore difficult to read. Even if one understands the rough connections fast, without a geological dictionary one will hardly be able to understand the details. Maps can be obtained at any time from local bookstores, but you can also contact the State Surveying Offices and State Geological Offices directly (if necessary, ask for a list of available maps). A corresponding internet research is an important basis for this.

Finally, it should be pointed out that literature is also helpful. For many landscapes in Germany, but also for interesting regions in other countries, there are comprehensible nature guides that include the geological developments. In addition, there are also numerous informative sources on the Internet.

Especially in cold-temperate areas, where many bedload findings are to be made, a descriptive rock identification book creates the prerequisites for the classification of such rocks (formation, origin, etc.).

For more intensive excavation work, a (folding) spade, an (old) knife, a folding rule, if necessary a geologist's hammer, a compass and a magnifying glass are useful. For topographical classification, it is helpful to determine the degree grid coordinates with the aid of a GPS device or the GPS function of a smartphone.

6.6.2   Variety of Shapes in a Confined Space: An Example

Figure 6.6 shows a schematically sketched section of a landscape shaped by the cold period. The Feldberger Seenlandschaft is located in southeastern Mecklenburg-Vorpommern, about 30 km southeast of Neubrandenburg. Its analysis lends itself to the fact that a variety of glacial and fluvioglacial forms can be found here in a very small area. This fact is due to the changeful geological history of the region. Thus, during the Vistula Cold Period, there have been warmer and colder periods due to noticeable changes in global temperature. During the colder phases, the ice sheet, coming from Scandinavia via the Baltic Sea basin, advanced towards north-central Europe; during the warmer phases, it temporarily melted back. The outermost ice margins of the three colder phases are marked by distinctive terminal moraines. These are the **Brandenburg Main Terminal Moraine**, the **Frankfurt Main Terminal Moraine** and the Pomeranian **Main Terminal Moraine**. The regional designation results from distinctive points or areas along this main terminal moraine, for example, the terminal moraine of the Frankfurt stage runs through the middle of the town of Frankfurt an der Oder.

As Fig. 6.6 makes clear, the Feldberg lake landscape now lies in the middle of the so-called Pomeranian main end moraine, i.e. the deposits that were deposited during the most recent main ice advance. The inland ice had reached a standstill here about 14,000 years ago. To the northeast of the main end moraine run further end moraine trains. For example, such an end moraine runs through the middle of the village of Feldberg, south of the Haussee lake, and parallel to it, about 2 km to the northeast, runs another end moraine, the so-called **Watzkendorfer Staffel**. Since such terminal moraines can only be built up when glaciers have at least a longer stagnation phase, it can be assumed that in this region alone three such stagnation phases have occurred. The lakes, the Haussee and the Breite Luzin, are embedded in the associated tongue basins, thus are tongue basin lakes.

Those who have the opportunity to look at the landscape on site will recognize the prominent elevation between the Breiter Luzin and the Haussee and will also hike the Reiherberg north of the Haussee. The comparatively high "dividing line" between the two lakes and the narrowing of the Haussee area from the northeast suggest that the terminal moraine of the Watzkendorfer Staffel was built up during a renewed ice advance and thus represents a compressive end moraine. Northwest of

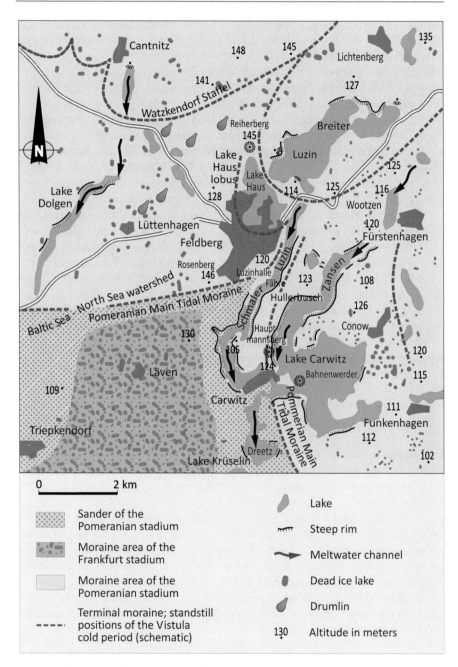

Fig. 6.6 Landscape shaped by the cold period in the area of the Feldberg lake district (© Wolfgang Fraedrich)

Feldberg several drumlins are arranged according to the main direction of ice advance. They are clearly visible in the real landscape and document that the inland ice has advanced repeatedly.

In the region immediately southeast of the present-day town of Feldberg, the former inland ice area has been drained. The Schmale Luzin, a lake clearly cut into the relief, is a typical gully lake, next to which two further gully systems run just under 2 km further east and around 6 km further west. They are all of subglacial origin, i.e. the meltwater has already run off under the ice in a tunnel valley. Lake Carwitz, on the other hand, was formed exaratively by the ice. Here, at the inner edge of the main terminal moraine, a lake formed in a depression after the melting of the inland ice and has survived to this day. In addition, numerous small lakes or water holes (sinks) mark the location of dead ice blocks towards the end of the ice age.

The formerly glaciated area alone shows the direct juxtaposition of glacial and fluvioglacial erosion and depositional forms. In the southwestern part of this landscape section, two things stand out. First, the fact that immediately in front of the former glacier gates (at the southern end of the Dolgener See, at the Schmaler Luzin and at the western side of the Carwitzer See) outwash plains can be found. Thus, a part of the material carried by the meltwaters has already been deposited here. Secondly, it is noticeable that this outwash area does not extend over the entire former glacier forefield. Rather, the elevations 1.5 km to the northeast (130 m) and 2.5 km to the southwest (119 m) of Läven mark a weakly developed ridge of elevation, which also has numerous troughs. Interpreting only via this sketch map is certainly the most difficult detail to tackle here. Two, questions arise:

- Why are there not outwash deposits throughout the area?
- If there is no outwash area in the central area, what is it?

Well, meltwater streams flow according to the largest gradient, and that was certainly given to the west and east of this ridge, which is at least 10 m higher. The elevation of 130 m, which is quite striking, acted at that time like a pillar around which the water had to flow.

As a result, the troughs have not been filled with meltwater sediments (sand and gravel), and this also provides an answer to the origin of this ridge. It is part of the ground moraine landscape, which was formed during the penultimate main ice advance, that of the Frankfurt stage, and in which dead ice blocks were preserved when the ice melted back, here about 15,000 years ago.

This again makes it clear that individual forms can be interpreted in terms of their origin, but that their relatively precise chronological classification is only possible by deciphering the system as a whole.

6.6.3   Form Society in the Model: The Glacial Series

Albrecht Penck (1858–1945), one of the most famous geographers of recent history, became known for his numerous works on glacial morphology. At the beginning of

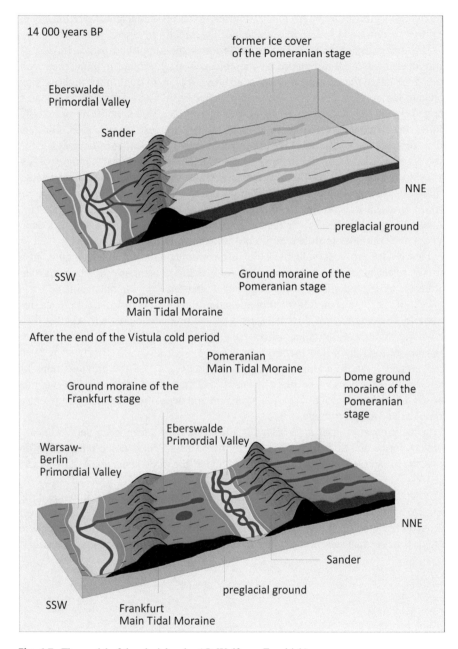

Fig. 6.7 The model of the glacial series (© Wolfgang Fraedrich)

the twentieth century, he carried out investigations in the Alpine and Alpine foothills region over many years in order to reconstruct the glacial history of the formation of the Alps. He coined the term glacial **series** (Fig. 6.7), which is used to explain the structure of a glacial and fluvio-glacial landscape. The essential features of the model will first be presented before its significance is critically reviewed.

The model-like aspect of the glacial series is the simplified causal linking of individual forms of areas shaped by the cold period. A ground moraine area with embedded lakes, the terminal moraine, the outwash plain in front of it – in the Alpine foreland the gravel plain – and the glacial valley form, so to speak, the unit that is formed within the framework of an ice advance cycle. This simplification initially allows the reduction of the complex facts to the simplest relationships: Glaciers advance, erode rock material at the base, grind it during transport and deposit it at the base (ground moraines) and in front of the glacier front (terminal moraine). They excavate the bedrock into hollow forms (tongue basins) and release meltwater, which erodes beneath the ice (meltwater channels with channel lakes) and deposits eroded material outside the ice margin – beyond the terminal moraine (outwash plains, gravel plains). The meltwater collects in receiving waters, e.g. a glacial valley, and continues to flow over it until it is eventually discharged into the sea.

This model provides valuable help in developing a basic understanding. In a strictly scientific sense, however, it is not tenable. Too many components that influence the evolutionary history of a glacial landscape remain unconsidered.

First of all, there is the so-called preglacial initial relief. It makes a difference, for example, whether glaciers flow at a steep gradient or a gentle gradient, whether they flow over an almost flat relief, whether they have to overcome obstacles, whether they incorporate them into the moraine debris carried along or whether they even flow around them. In addition, fluvio-glacial processes – as the previous remarks have made clear – can take place in many different ways, depending on various factors that indirectly control erosion, transport and deposition and thus give rise to a complex landscape structure.

In the present, glacial geologists not only study the Pleistocene and recent (= present) forms and their formation, but they also try to get detailed information on how the processes in (intraglacial) and under the ice (subglacial) take place, i.e. today they analyse more the dynamics of the system.

It is hoped that this will provide more differentiated insights into the flow behaviour of the ice and meltwater, which – together with the glacier level measurements of recent mountain and inland ice glaciations – will also enable forecasts to be made about the future development of today's ice margin areas.

Shape Design Away from Glaciated Areas

7

Wolfgang Fraedrich

Abstract

Glacial ice and meltwater are important exogenous forces in the polar and subpolar regions of the Earth, but also in the high altitudes of the high mountains. The transition zone between the glaciated regions and the (cool) temperate latitudes is characterised by a subpolar climate, in which permafrost also occurs. These are in large parts so-called periglacial areas, regions which – often in company with fluvioglacial forms – show a specific periglacial form. This chapter provides information about

- the phenomenon "periglacial" itself,
- the framework conditions for the development of periglacial relief dynamics,
- the regional distribution of periglacial areas,
- the specific periglacial formations (aeolian, gravitational, cryogenic).

It has been pointed out several times that relief-forming forces and processes are also effective away from the glaciated areas. They are specifically controlled by the subpolar climate, which is characterized by long, cold winters as well as short, cool summers and transitional seasons lasting only a few weeks. According to current knowledge, it can be assumed that these climatic conditions can also be transferred to the Pleistocene ice margins, the **periglacial regions**.

▶ **Periglacial** Periglacial (from *peri* = at the edge, and *glacial* = shaped by the ice) means at the "edge of the ice". This word is used as an adjective regionally, climatologically and morphologically. Periglacial areas are thus those areas in

W. Fraedrich (✉)
Hamburg, Hamburg, Hamburg, Germany
e-mail: wolfgang.fraedrich@hamburg.de

© The Author(s), under exclusive license to Springer-Verlag GmbH, DE, part of
Springer Nature 2023
W. Fraedrich, *Traces of the Ice Age*, https://doi.org/10.1007/978-3-662-65886-4_7

front of the glaciated area that are directly and indirectly influenced by glaciation. They exhibit a typical periglacial climate, a subarctic or even subpolar climate, which in turn brings with it typical morphodynamics and thus a typical set of forms.

7.1   Overview of Forces and Processes

Three factors play an essential role in the design of the shape:

- Due to the short growing season, the vegetation is only very sparsely distributed, although it becomes increasingly dense and taller from the ice edge until it finally changes into forest lands (e.g. the taiga) outside the periglacial areas.
- The long-lasting frost, the consequence of which is the deep freezing of the soil, one speaks here of permafrost soil (= permafrost soil, lat. *permanere* = to remain).
- Outside of winter, there are many so-called frost change days, i.e. days on which the temperature drops below freezing point at night and rises above 0 °C during the day.

As a result, various processes play a role in landscape formation. In the context of the morphodynamic cycle, mechanical weathering must be mentioned first, and here in particular frost alteration weathering or frost blasting. It prepares the rock, breaks it up or crushes it, before removal, transport and deposition follow. Transport or mass displacement is achieved by **aeolian processes** (Greek *Aeolus* = god of the winds), gravitational **processes** (Latin *gravitas* = gravity) and **cryogenic processes** (Greek *kryos* = cold, frost). The wind, gravity and frost are therefore the controlling forces.

7.2   The Regional Distribution of Periglacial Areas

Figure 7.1 shows the zonal distribution of recent periglacial areas in the northern hemisphere. Of course, periglacial processes also play a role in the subpolar regions of the southern hemisphere (e.g. at the edge of Antarctica) and in the so-called *subnival* (lat. *nivalis* = snow) stages of the high mountains, where the typical frost-change climate dominates.

Geologists differentiate between a zone of continuous permafrost and a zone of sporadic permafrost. The map shows that the zone of continuous permafrost is located near the pole, while the zone of sporadic permafrost adjoins it to the south and even extends to almost 45° north latitude in Central Asia. Corresponding to the latitude, the thickness of the **permafrost** decreases from the pole southwards. Figure 7.2 shows a north-south vertical profile across Canada, from which it can be seen that at Resolute, a small climate station at 74° north latitude, the thickness of the permafrost soil is more than 430 m. At Norman Wells, which is still a good 800° north latitude, the thickness of the permafrost soil decreases from the pole

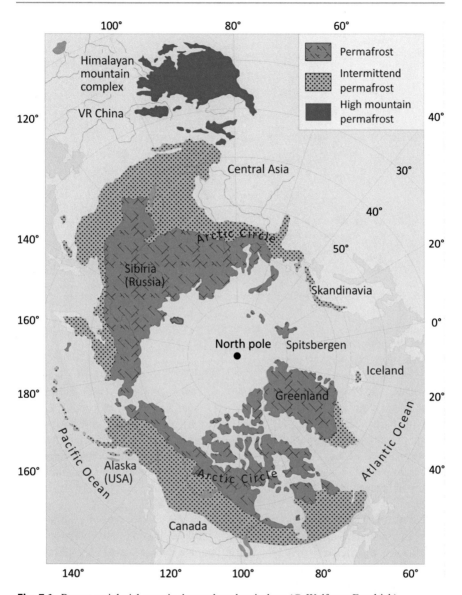

Fig. 7.1 Recent periglacial areas in the northern hemisphere (© Wolfgang Fraedrich)

southwards. At Norman Wells, a good 800 km further south at latitude 65°N, the transition from continuous to sporadic permafrost occurs; here the thickness of the latter is still up to 50 m.

The thickness of the summer thaw **layer develops in the** opposite direction. At Resolute, the ground thaws to a depth of only 0.15 m, whereas in the summers at Norman Wells, which are several weeks longer, it thaws to an average depth of just under 1.70 m. The term "active layer" used for the thawing layer in the Anglo-Saxon

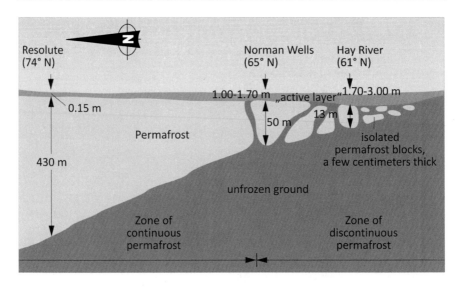

Fig. 7.2 North-south vertical profile of permafrost in Canada (© Wolfgang Fraedrich)

language makes it clear that part of the morphodynamic processes take place here, namely the gravitational and cryogenic processes.

During the cold periods, the periglacial areas, viewed globally, extended further south in the northern hemisphere, correspondingly further north in the southern hemisphere, and in the high mountains to lower altitudes. Even during the Weichselian/Wurmian glaciation, when only Scandinavia, north-central Europe and large parts of the British Isles were covered by the ice sheet and the Alps by a plateau glaciation, there were no forest landscapes in the whole of western, central, eastern and parts of south-eastern Europe. The area including the southern North Sea basin was covered by a sparse tundra, which was completely absent in the immediate vicinity of the ice margin and probably also in the high altitudes of the low mountain ranges.

As a result of the climate zone shift at that time, the ice-free area of Europe north of the Pyrenees and the Alps was subarctic until far into the east of Europe and consequently a periglacial area. One has to imagine the climate as we find it in the present in those periglacial areas that are delimited on Fig. 7.1. This also explains why we find traces of Pleistocene periglacial structures all over Germany.

7.3   The Periglacial Form Treasure Trove

In the following, periglacial morphodynamics are explained using various examples. Here, too, the differentiation is made according to the effective forces and processes. In doing so, it is to be made clear, with recourse to recent as well as Pleistocene structures, that the underlying regularities run or have run independently of developments in the history of the earth. The structures and processes observed in recent phenomena are therefore to be transferred to prehistoric ones, including Pleistocene ones (principle of actualism).

7.3.1 Wind: Aeolian Processes and Forms

"Driving through the Icelandic highlands in summer. In the middle of the highlands was once again a break for the night. Already in the evening wind came up gradually, but still the sun shone far into the subpolar summer night. Our small tents were already battered by the wind in the early morning. Dismantling proved difficult; two of us could hardly manage it, for the tarpaulins were torn from our hands by the force of the wind. The most necessary things were done, the luggage was stowed in the vehicle, then the journey continued. The sky was black-grey in the meantime, then the cloud cover loosened up for a short time, only to be jet black again shortly afterwards. The Langjökull was not to be seen any more. Not because it was raining or the clouds were too low. No, finest dust was whirled through the air: dust storm on Iceland!"

This little report can only approximate the experience. Dust storm – why at all, and what are the consequences of such storms?

The Icelandic highlands belong to those regions of the earth that are shaped to a very decisive extent by periglacial processes. The island, which is built up of predominantly basaltic rock of young volcanism – Iceland is part of the Mid-Atlantic Ridge, a seam in our solid earth's crust – has repeatedly been completely glaciated during the last cold periods. The highland area to the west is no longer in the zone of active volcanism. Here the traces of the recent glaciation during the Weichselian cold period are apparent. Detersive, detractive and exarative moraine material is present on the surface. Ecologically "desolate" stone deserts (isl. *hraun*) often characterize the picture here. There is seldom a plant, no trace of dense vegetation. The summers are too short, the conditions for soil formation too unfavourable in this hostile periglacial climate. The chemical weathering processes take far too long to form nutrient-binding clay minerals, and the few fertile soil components are usually blown away by the winds, which transport them over many, many kilometres before depositing them elsewhere. And this happens several times during a short summer of about 3–4 months.

As the Icelandic highlands appear to us today, we must imagine large parts of Central and Western Europe during the Weichselian/Würmian cold period. A landscape poor in vegetation, in which the violent winds – mostly coming from the southwest to northwest – prevailed at the earth's surface, even at that time. Over thousands of years, millions of tons of fine material were transported by the wind during the short snow-free summers. Wherever obstacles stood in the way of the "air freight", e.g. at the northern edge of the low mountain range zone, in the **Börden**, large quantities of this fine sediment were deposited again. A smaller part, however, was also transported into the low mountain range region and was then deposited in basin landscapes as far south as southern Germany. Geologists refer to this aeolian fine sediment as loess.

Fig. 7.3 Dune sands near Tornesch (Schleswig-Holstein) from the transitional phase from the Late Glacial to the Early Holocene (© Wolfgang Fraedrich)

▶ **Borders** Börden are the bay-like foothills of the North German Lowlands (e.g. Jülicher Börde, Soester Börde, Hildesheimer Börde, Magdeburger Börde). Here, fertile soils, the loess loams, have developed on loess deposits, some of which are several metres thick, enabling the most intensive agricultural use (sugar beet and wheat cultivation predominate).

▶ **Loess** Loess (in Swiss "lösch" = loose) is a yellow to yellow-brown aeolian dust sediment with a predominant grain diameter of 0.01–0.05 mm. Mineralogically, the main constituents are quartz grains and also lime particles. The lime content of the loess can be up to 20%. The Pleistocene loesses originate to a large extent from the outwash plains and gravel terraces of the rivers of that time. The chemical

Fig. 7.4 Dune sands near Neu Wustrow (Mecklenburg-Western Pomerania) from the High Glacial period (© Wolfgang Fraedrich)

weathering of the limestones triggered by the penetration of carbon dioxide-rich seepage water leads to the formation of the loess loams.

In addition to loess, however, sands have also been moved aeolian. In numerous regions of the Pleistocene periglacial area of Europe there are fossil drift sand and dune deposits. Figures 7.3 and 7.4 show dune sand profiles, the first from the transitional phase from the Late Glacial to the Early Holocene and the second from the **High Glacial** period. The emplacement of the profile from near Tornesch (Fig. 7.3) shows interbedded humus horizons a few centimetres thick, representing soil formation processes from the Bölling Warm Period (c. 13,300–13,000 years before present, middle) and the Alleröd Warm Period (c. 12,500–12,000 years before present, top) (Figs. 4.6 and 4.7). This has been confirmed by C14 dating and also pollen analytical studies. The other profile (Fig. 7.4) shows a very fine stratification, which is particularly typical for eolian deposits. Due to the different resistivities of the material, the structures are strikingly carved out. Microscopic **grain shape analysis** and **sieve analysis make** it possible to determine dune sands like these.

▶ **Grain Shape Analysis** Grain shape analysis aims to determine the degree of rounding of unconsolidated sediments. Predominantly rounded particles have been subjected to high mechanical stress during transport, i.e. have experienced strong abrasion. This is the case both during transport by water and transport by wind.

▶ **Sieve Analysis** Sieve analysis is used to separate the individual grain size fractions from each other. Sieves with standardized mesh sizes corresponding to the grain size classification from coarse gravel to medium gravel, fine gravel, coarse, medium and fine sand to silt and clay enable the sieving of a dried sediment sample. Since the sediments moved and deposited by different transport forces have very specific grain size compositions, the classification of a sieve sample result is possible with the help of the determination key. For dune sands, a very clear predominance of fine sand (0.063–0.2 mm grain diameter) with an average share of up to 70% and a high share of medium sand (0.2–0.63 mm grain diameter) with a share of almost 30% is found.

Practical analyses of samples from both outcrops have confirmed their interpretation as dune sands. The fine sands dominate – in the absence of coarser particles – and are predominantly rounded.

7.3.2   Gravity: Gravitational Processes and Forms

Gravitational processes, i.e. processes triggered and controlled by gravity, are not typical of periglacial regions alone. For example, landslides after heavy rainfall are also possible in the tropics and in temperate latitudes. In the subpolar climate regions, however, they play a significant role.

Relief formation by gravitational processes occurs essentially without transporting agents, merely by the action of gravity on mobile rock masses. In the periglacial region, freezing and thawing processes also have a decisive effect and control the corresponding mass movements. In addition to this regional or climate-specific aspect, there are other influencing factors, such as the type of rock (including bedding, structure, water permeability, water content), the strength of the weathered solid rock and the cohesiveness of unconsolidated rocks to tensile and compressive forces, and finally also the relief shape.

Since there is usually little vegetation in the periglacial areas, the soil lacks firmness, and there is no possibility of releasing soil water back into the atmosphere via the (albeit small) evaporation activity of the plants. Now, when the soil begins to thaw from above after the end of winter, it absorbs water until it is saturated. The degree of saturation is usually reached in a few days, regardless of the pore size of the substrate, because

Fig. 7.5 Flowing soil overlying meltwater sands in a gravel pit in the Harburg Hills, northern Lower Saxony (© Wolfgang Fraedrich)

- a large amount of water is released by the onset of snowmelt; although some of this runs off superficially, it also passes into the soil water via seepage;
- seepage of the soil water into greater depths is not possible, as the soil only thaws on the surface.

The thawing layer, which is thus saturated with water after a short time, begins to flow depending on the slope inclination. This process is called **solifluction**.

▶ **Solifluction** Solifluction (Latin *solum* = the soil and *fluere* = to flow) means soil flow, a mostly slow flow process, which sometimes reaches only a few millimetres per day, but with a corresponding slope inclination also more than 1 m per day. Depending on the vegetation cover, a distinction is made between the more intensive free solifluction in the frost protection zones and the slower bound solifluction in the tundras. Slopes as steep as 2° are sufficient to set the tough, water-saturated "soil mush" in motion and thus create typical flowing earth structures – usually over many years.

Numerous sand and gravel pits in Central Europe show so-called flowing soils in the outcrop profile. In the area of glacial deposits, the **boulder clay and boulder loam of** the ground and terminal moraines were transported away, whereby a levelling of the glacial relief took place, so that terminal moraines were lowered and, above all, dome-shaped ground moraines were removed superficially. Partly the clayey material has also flowed over fluvioglacial sediments. Frozen sand lenses may have been included in the flow process and moved along for several tens of metres (Fig. 7.5).

Flowing soils show neither a stratification nor a clear material sorting of another kind in their structure. They differ from original moraine material, however, due to the stronger intermixing with gravels, which are also predominantly bedded according to the original flow direction.

7.3.3 Frost: Cryogenic Processes and Forms

In the frost protection zones of periglacial regions, one encounters in many places "mysterious stripes and rings" of stones that form a pattern on the ground surface (Fig. 7.6). Stones of approximately the same size are lined up as if laid out by human hand. If they lie on a flat surface, they are almost circular stone rings, on slightly sloping surfaces (slope inclination about 2–2.5°) they are stone ovals, on larger slope inclinations (from about 6°) they are stone strips, which run according to the greatest slope. Such structures are found today only in recent periglacial areas. To explain the phenomenon, two aspects have to be considered:

- the interplay of **frost heave** and **frost pressure in** the thaw layer of the permafrost area,
- the bottom flowing already described before.

Fig. 7.6 Structural soil with stone ovals in the Icelandic highlands (© Wolfgang Fraedrich)

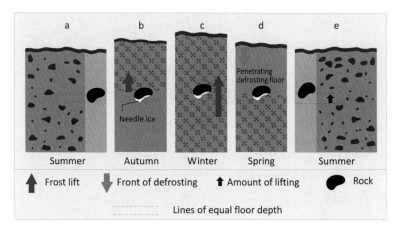

Fig. 7.7 The principle of frost heave and frost pressure (© Wolfgang Fraedrich)

Frost heave and frost pressure cause sorting of the weathered rock material. The seasonal temperature change that takes place in the thawing layer sets in motion a movement of the rocks as shown in graphically simplified form in Fig. 7.7. Phase a shows the initial position of a larger rock in a thawing layer in summer. With the onset of frost at the beginning of autumn (phase b), the volume of the thaw layer saturated with water increases due to the expansion of the water in the frozen state. Since it is not possible to move downwards (permafrost soil), the soil expands upwards. The stone is lifted in the process. As the freezing front advances into the depths, the cold penetrates the rock faster than the surrounding sediment, causing small ice crystals to form on its underside (= needle ice, phase b).

The freezing front finally advances to the upper limit of the permafrost (phase c). When thawing begins again in early spring (phase d), fine material first sinks to the place of the faster-thawing needle ice, so that the stone, although the material loses volume again, now lies embedded higher in the sediment than in the previous year (phase e). This process repeats itself from year to year. Incidentally, this is a process that can even be demonstrated in temperate climates, namely where fine-grained soils are present at shallow depths on solid rock substrates, in which rock remains are moved from the base to the surface by the frost heave.

A slightly more complicated but more real representation of structural soil formation is shown in Fig. 7.8. The four phases of **structural soil formation** show an annual cycle, and the thaw layer is shown in longitudinal section. Here it is clear that coarse material is separated from fine material by frost heave and frost pressure. With the onset of cold weather in autumn, the thaw layer freezes from above into the depths. Pressure is exerted on the still soft subsoil of the thaw layer via the upper frost front, and the permafrost layer generates pressure from below. Between these two solid fronts, the remaining thaw layer is compressed, and the pressure is released both upwards (bulging of the fine soil) and to the sides (compression of the coarse debris). The stones frozen out by the frost heave slide

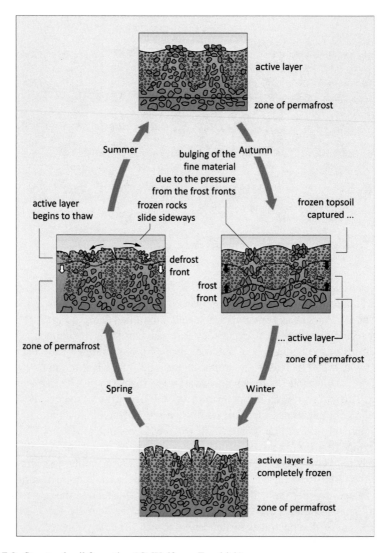

Fig. 7.8 Structural soil formation (© Wolfgang Fraedrich)

sideways on the curved (convex) surface when thawing begins in spring, thus reinforcing the stone ring. If such a process now takes place on flat or slightly inclined surfaces, these distinctive stone patterns (structural soils, polygonal soils) form, as shown in Figs. 7.6 and 7.9. Geologists refer to processes of this kind as **cryoturbation**.

▶ **Cryoturbation** Cryoturbation (Gr. *kryos* = frost, cold; Lat. *turbare* = swirling) comprises the sorting processes of soil and rock material of different grain size in a

Fig. 7.9 Structural soil with stone strips on the plateau of Snæfellsjökull volcano in West Iceland (© Wolfgang Fraedrich)

periglacial thaw layer, triggered and controlled by soil frost. Cryoturbation phenomena can be extremely diverse. However, they can all be traced back to the basic mechanical principles of frost heave and frost pressure associated with the seasonal frost change, under the influence of solifluction. The flow earth structure on Fig. 7.5 was also exposed to these mechanisms; it is a witness to Pleistocene periglacial dynamics.

If, for example, sand layers came under the periglacial influence, permafrost was formed in the subsoil and the thawing layer on the surface. Year after year, the thawing layer was able to flow, usually with only a slight slope inclination, so that especially the thin layers built up from silts and clays in an outcrop wall no longer show an ordinary layer structure, but – depending on the intensity of the cryoturbation – run in a curved manner.

Very striking and impressive periglacial structures are **ice wedges**. Such structures on the surface are found in recent periglacial areas only with very diffuse tundra vegetation. Here, ice wedge nets running at depth can be seen on the surface, sometimes regularly, sometimes irregularly. The starting point of ice wedge formation is a rupture of crevasses in the frost soil at temperatures below 0 °C, the so-called **deep frost shrinkage**.

Fig. 7.10 Syngenetic ice wedge in a gravel pit in the Harburg Hills near Hamburg (© Wolfgang Fraedrich)

▶ **Deep Freeze** The phenomenon of deep frost shrinkage (in Anglo-Saxon usage, the more expressive terms "thermal contraction" or "frost cracking" are common) is based on the physical property that ice – like any solid body – contracts again when it cools significantly below 0 °C, by about 0.05 mm per metre of ice per °C. This means that over a distance of 1 m at the ground surface, a gap of 1 mm opens up. On a distance of 1 m at the ground surface, therefore, a gap of 1 mm tears open at −20 ° C. The decisive factor here, however, is not the absolute thickness of the ice. However, the decisive factor is not the absolute depth of the temperature, but the rate of temperature change per unit of time. Another parameter is the dependence on the strength of the material. A temperature drop of as little as 4 °C will result in a crack in a very icy soil, while a temperature drop of at least 10 °C is required in a rocky material. The so-called initial crack can reach down to a depth of 4 m. In the thawing layer it is completely destroyed again with the onset of spring. However, if it has reached the permafrost zone, it will start there the next year.

It takes many years before larger ice wedges can form. Due to the physical conditions, active ice wedge formation is only possible where annual mean temperatures of −6 to −8 °C are reached, i.e. in zones of continuous permafrost. In areas of sporadic permafrost, the thaw layer is too thick.

Figure 7.10 shows a fossil ice wedge. The meltwater sediments were exposed to periglacial climatic conditions during the end of the Saale glacial period. The sediment body shows a clear stratification in which partly coarser sediment sequences alternate with finer ones. At the edge of the wedge the layers are drawn into the depth, so it is an ice wedge, which was formed during the deposition of the sediments (= **syngenetic ice wedge**).

Once a deep frost shrinkage gap has formed, it can no longer close under constant climatic conditions. This is explained by various factors acting together. In many cases, such contraction gaps are already covered with sand at the onset of winter, and snow is deposited during the winter. When the summer thawing phase begins, hoar frost forms in the crevice due to the higher outside temperature, and finally meltwater accumulates in the crevice as it continues to thaw. Fine sediments are transported along and gradually form a recognisable wedge filling. In the coming autumn, the process begins anew, and it repeats itself from year to year. The ice wedge becomes wider and wider and grows further into the depths. Another decisive factor is the fact that the ice wedge filling thaws faster than the neighbouring rock due to the composition of the material. The higher temperature causes the thawing ridge to move from the wedge to the neighbouring rock, where the water bound in the form of ice is thawed and eventually extracted (**dehydration**). As frost resumes, the water-saturated wedge fill gains volume, the wedge widens, and so on. In addition to the deep-freeze shrinkage theory, the dehydration theory is also used to explain this.

In contrast to the syngenetic ice wedge, there is also the **epigenetic ice wedge**, which develops after rock formation and in which the edges of the wedge are curved upwards (Fig. 7.11).

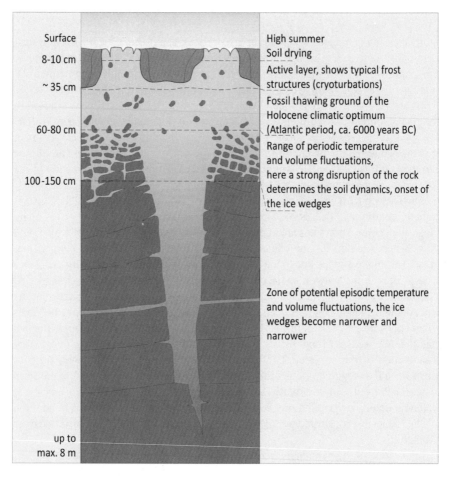

Surface
8-10 cm

~ 35 cm

60-80 cm

100-150 cm

up to
max. 8 m

High summer
Soil drying
Active layer, shows typical frost
structures (cryoturbations)
Fossil thawing ground of the
Holocene climatic optimum
(Atlantic period, ca. 6000 years BC)
Range of periodic temperature
and volume fluctuations,
here a strong disruption of the rock
determines the soil dynamics, onset of
the ice wedges

Zone of potential episodic temperature
and volume fluctuations, the ice
wedges become narrower and
narrower

Fig. 7.11 Graphical representation of a recent ice wedge structure in southeast Spitsbergen (© Wolfgang Fraedrich)

Glaciers and Sea Level

8

Wolfgang Fraedrich

Abstract

If you read current media reports or follow thematically relevant documentaries, the connection between glaciers and sea level is emphasized again and again. This is understandable, because melting glaciers release freshwater, a considerable proportion of which enters the world's oceans via surface runoff – so sea levels will rise if glacier retreat continues worldwide. However, the development of the North Sea and the Baltic Sea is one example that shows that this is not only a modern process.

The reader will learn more about this in this chapter.

- On the one hand, it is the principle of isostasy, or more precisely the physical principle of isostatic equilibrium, which has shaped the marine situation in the Baltic Sea region in the course of the Earth's recent history. The Nordic ice sheet repeatedly pushed down the earth's crust with its uplift; if the ice sheet masses melted, the land could rise again.
- In addition, there is the principle of eustatic sea-level rise mentioned at the beginning – when ice melts, fresh water enters the world's oceans, and sea level rises.
- These two principles work together, this is explained by the example of the post-glacial development history of the Baltic Sea region.
- The geological structure of the marshes along the German North Sea coast is used as an example to show what traces of sea-level fluctuations can be read from them.

W. Fraedrich (✉)
Hamburg, Hamburg, Hamburg, Germany
e-mail: wolfgang.fraedrich@hamburg.de

- In a final comparative consideration of the North Sea coast and the Baltic Sea coast, it is explained how these landscapes can be grasped in terms of their complexity.

At the narrowest point of the Gulf of Bothnia between Sweden and Finland, new islands have been emerging again and again for millennia, and the existing islands

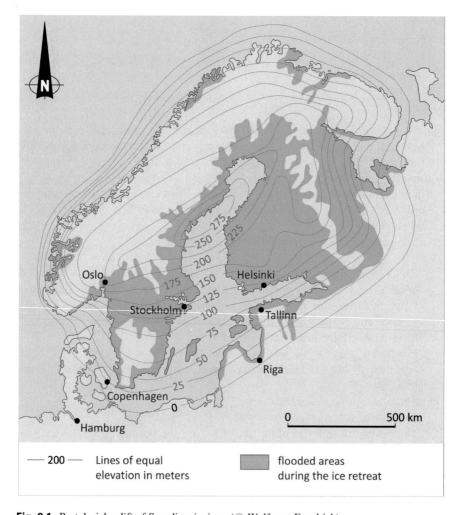

Fig. 8.1 Postglacial uplift of Scandinavia, in m (© Wolfgang Fraedrich)

have been increasing their area, which has led, for example, to the Finnish landowners of the Raippaluoto archipelago distributing the newly created land – about 500 km^2 each time – every 50 years. The land emerges from the sea (Fig. 8.1). How can this be explained?

8.1   The Geophysical Principle of Isostatic Movements

In order to interpret the phenomenon described above, one must go back about 12,000 years in the history of the earth. Towards the end of the Weichselian, the land masses of Norway, Sweden and Finland, the entire Baltic Sea basin and the northern edge of Central Europe lay under a mighty ice sheet, the greatest thickness of which in the area of the Gulf of Bothnia was about 2500 m. The weight of this enormous ice sheet weighed down the earth's crust, which "floated" on the viscoplastic mantle and was consequently pushed down into the depths. Then, after the ice gradually melted back, the surcharge decreased again, and the earth's crust has since resurfaced.

Such vertical crustal movements are, besides the horizontal crustal movements, typical expressions of endogenous (= internal) forces. The vertical movements have very different causes; what they all have in common is that they – like the movement of the lithosphere plates – proceed very slowly. They are millimetre amounts per year (Fig. 8.2), really noticeable dimensions can only be observed in geological periods.

In order to be able to explain these vertical movements at all, one must realize that the **asthenosphere** is in a viscoplastic state.

However, the upper mantle rock is by far not molten, as one reads and hears everywhere, but it is very viscous (= viscous). It is kept in motion by the heat released from the earth's interior, so that it can move by a few centimetres per year. This viscosity is also the prerequisite for individual crustal floes to also move vertically.

Continental and oceanic crusts have different **densities**, with the continental crust having an average **density** of 2.85 g/cm^3, whereas the oceanic crust has an average **density** of 3.31 g/cm^3. By nature, a floating equilibrium (= **isostasy**) should be established due to the physical laws, which does not succeed de facto due to the ongoing endogenous dynamics, but also due to the exogenous processes.

▶ **Asthenosphere** The asthenosphere is a zone of low material strength in the Earth's uppermost mantle, extending from a depth of about 100 to 300 km. The rigid earth's crust, which is divided into individual plates, "floats" on the viscoplastic material.

▶ **Density** The density (= specific gravity) of a material results from the ratio of mass (given in g or kg) and volume (given in cm^3 or m^3): $\rho = m/V$ with the unit g/cm^3 or = kg/m^3.

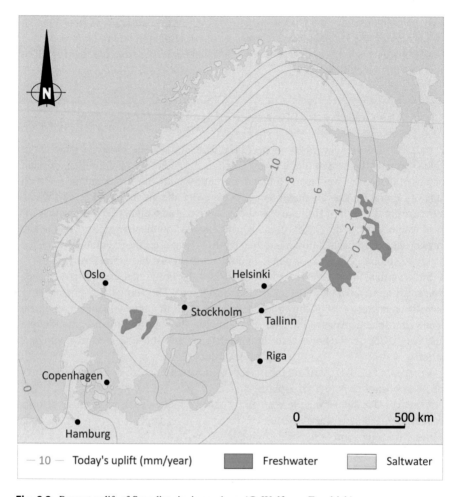

Fig. 8.2 Present uplift of Scandinavia, in mm/year (© Wolfgang Fraedrich)

▶ **Isostasy** Geologists use the term isostasy (from Greek "*isos*" = equal in number, size, strength, significance, etc. and "*stasis*" = standing) to describe the mass equilibrium within the Earth's crust system and thus a geological state of equilibrium between the Earth's crust and the mantle below it. Disturbances of the isostatic equilibrium are compensated by crustal parts of different density and weight rising or sinking until a floating equilibrium is established. Thus, depending on their gravity, they plunge deeper or less deeply – in proportion to their thickness – into the asthenosphere. The British astronomer and physicist George Airy had already developed the theory of isostasy in 1855.

Glacial isostatic uplift processes are detectable throughout the Baltic Sea region, but also along the Norwegian and Icelandic coasts. If, for example, you take a ferry

Fig. 8.3 Fossil surf cave on a postglacial beach terrace in the south of the island of Gotland (© Wolfgang Fraedrich)

to Gotland, you will notice long before the ship enters the harbour that the town of Visby, which was already important in the Middle Ages, has been built on various beach levels that cannot be explained by the erosive force of the sea. Figure 8.3 documents the phenomenon by a surf cavern that is today about 4 m above sea level.

8.2   Eustatic Sea Level Fluctuations

It has already been mentioned that the area of the southern North Sea was periglacial during the Weichselian period, i.e. it was neither glaciated nor – as today – covered by water. About 20,000 years ago, when the inland ice reached its greatest extent during the Brandenburg stage, the sea level worldwide was more than 130 m lower than today, about 17,000 years ago it was still −95 m, 11,700 years ago, during the younger tundra period (= younger Dryas period) and thus already in the Holocene, still −36 m. Only 8600 years ago Great Britain was separated from the mainland again, when the sea arms advancing from the north and the south united at about the height of the Dutch island Texel. On average, the sea level had risen by 2 cm/year by this time. After another 5000 years of somewhat slower sea level rise, the line of today's mean high water was reached.

Today, it is assumed that the sea level was up to 3 m higher than at present even after the beginning of our era. The extensive marshlands, which today are only just above sea level, as a result of marine and perimarine sedimentation (peri = at the edge, perimarin = at the edge of the sea) are cited as evidence for this assumption.

The cause of this global sea level rise is easily found. The millions of cubic kilometres of inland ice and high mountain glacier ice that had formed worldwide during the Weichselian/Wurmian cold period gradually melted back to the glacier areas that exist today with the gradual rise in global temperature and were fed as fresh water to the world's oceans. Sea level rise was the logical consequence. Geologists refer to the lowering of sea level caused by the binding of water in the form of ice as well as the rising sea level caused by the release of water when the ice melts as an **eustatic process**.

▶ **Eustasie** Eustasy (= eustatic sea-level fluctuations) refers to simultaneous, global and long-term changes in sea-level (positive = transgression; negative = regression).

These eustatic sea level fluctuations played a role for humans from the time they began to settle coastal areas. The first settlement of the marshes already before the birth of Christ had to be abandoned again when a short-lived transgression occurred. The building of the mounds (Warften) was a reaction of the people to this change of the sea level. A permanent settlement of the North Sea coastal area, however, only became possible with the construction of dikes (from about 1000 AD).

Although man is building much higher and more effective dikes today than in the past, future settlement of the coastal zone is at risk. Further sea-level rise would presumably – according to forecasts – be capable of triggering major flood disasters year after year with more frequent storm surges. This imminent danger is prompting the countries bordering the North Sea, especially the Netherlands and Germany, to invest more in coastal protection. The hypotheses discussed in this context will be dealt with in more detail in Chap. 9 (Fig. 8.4).

8.3   The Geological Structure of the Marshes Provides the Key to Reconstructing Sea-Level Changes

An outsider naturally wonders how it is possible to make such concrete statements about the temporal sequence of eustatic sea-level changes. It is true that people already lived in Central Europe during the Vistula Cold Period, but the sea-level rise took place far away from their habitats at that time. Even if they had been aware of it,

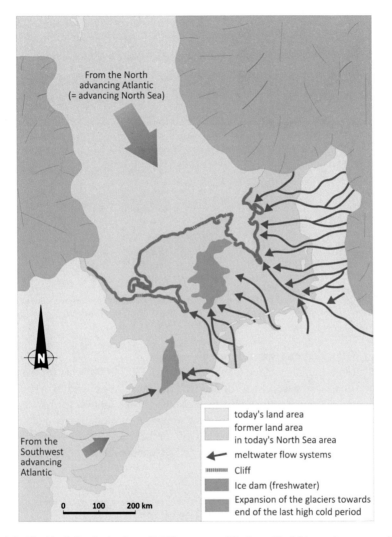

From the North
advancing Atlantic
(= advancing North Sea)

From the
Southwest
advancing
Atlantic

today's land area

former land area
in today's North Sea area

meltwater flow systems

Cliff

Ice dam (freshwater)

Expansion of the glaciers towards
end of the last high cold period

0   100   200 km

Fig. 8.4 The North Sea basin about 12,000 years ago. The Late Glacial is coming to an end, the North Sea is already advancing (© Wolfgang Fraedrich)

science today would have no records of that time available, people were not that far along at that time.

Here, as in many other research projects with similar objectives, the geological subsoil has been investigated more closely. With the aid of boreholes, information was obtained about the vertical sequence of the sediments. Figure 8.5 shows – in a very simplified way – the **stratigraphy** of the marshes in the East Frisian area. Thus, the peats at the various depths below the present sea level document a temporary standstill of the post-cold-age **transgression**, so that fen peat could form there over a period of several decades. **Pollen analyses with evidence of** typical fen vegetation

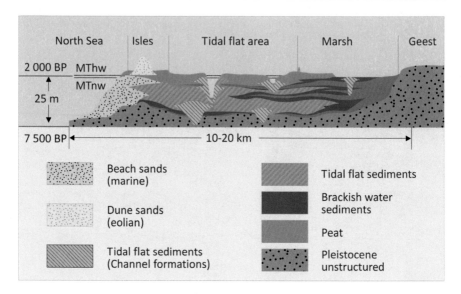

Fig. 8.5 Simplified geological profile from the East Frisian island of Norderney to the geological margin (*MThw* Mean Tidal High Water, *MTnw* Mean Tidal Low Water) (© Wolfgang Fraedrich)

(including alder and birch) have shown that this horizon of organic material is fossil fen peat. C14 dating has yielded an age of approx. 7500–2000 years for the peat.

Today's coastal landscape was formed by a rise in sea level of 25 m over a period of 5500 years. During this process, the coastline shifted an average of 10–20 km inland compared to today. An alternating sequence of peats (as a document of regressive phases) (**regression** = retreat of the sea) and marine deposits (as a document of transgressions) lies above the repeatedly flooded geest landscape, which was originally formed during the Saale cold period.

8.4 Isostasy and Eustasy Act Together

It seems plausible that the two processes described, isostasy and eustasy, do not occur independently of each other. If inland ice melts, the sea level rises, and at the same time the load of the ice body on the earth's crust sinks. This has been the case since the end of the Pleistocene. The interaction of the two processes will be illustrated in the following using the development of the Baltic Sea region.

Already 12,000 years ago, i.e. about 2000 years after the end of the last main ice advance (Pomeranian stage), the southern ice margin of the Nordic ice sheet in Europe had retreated by a good 600 km. In the area of the southern Baltic Sea basin, a large part of the meltwater collected in the **Baltic Ice Reservoir** (Figs. 8.6 and 8.7), i.e. an inland lake with fresh water. At that time, the sea level was so low (−90 m) that what is now southern Sweden was still connected to the mainland. In the central

Fig. 8.6 Icebergs in the glacier reservoir Jökulsárlon in southern Iceland – a similar picture will have existed in the Baltic Ice Reservoir at that time (although it is much larger) (© Wolfgang Fraedrich)

Baltic Sea, the same spectacle could have been observed as today off the coasts of Greenland or Antarctica. Here, the still flowing inland ice calved directly into the Baltic Ice Reservoir. The result was the formation of numerous icebergs.

The water level of the Baltic Ice Reservoir rose until the Middle Swedish Terminal Moraine was reached. The southwestern shore formed the Darss Sill, today a sea depth of −15 m in the southern Baltic Sea. At this time, the island of Gotland had not yet risen from the sea. The water level was a maximum of 28 m above the world sea level at that time.

About 10,000 BC, as a result of further melting of the inland ice, a connection between the Baltic Ice Reservoir and the world sea occurred at the Billingen Gate (between the present-day Swedish lakes Vänern and Vättern); the water level of the Baltic Ice Reservoir adjusted to the world sea level (at that time about −65 m). The salt-loving mussel *Yoldia arctica,* a leading fossil of the time, documents the salinity of the Baltic Sea at that time, the so-called **Yoldia Sea** (Fig. 8.8). Due to the overall still low world sea level, today's Denmark was completely mainland, there even existed a connection with the southern tip of today's Sweden.

In the meantime, the Scandinavian mainland had isostatically risen so far that it could "catch up" with the somewhat slower eustatic sea level rise. In central Sweden the connection of the Yoldia Sea to the world sea was interrupted, a renewed sweetening of the former Baltic Sea gradually began, in which Gotland now also

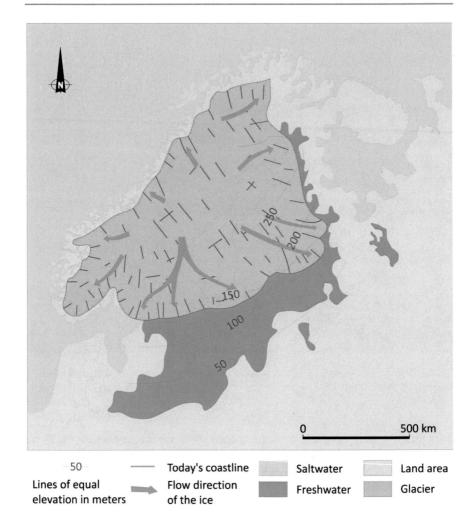

Fig. 8.7 Baltic ice dam (10,000–8000 BC) (© Wolfgang Fraedrich)

already existed as an island. The snail *Ancylus fluvialis,* which was found as a leading fossil in sediments, gave its name to the then **Ancylus Sea** (Fig. 8.9). The water level rose again to 14 m above the world sea level, but at −16 m it was still below the present value.

With the further melting back of the inland ice, the Ancylus Sea soon reached the Darss Threshold, there was again an access to the world sea and an adjustment to the world sea level. In the course of this Littorina transgression (after the snail *Littorina littorea*), salt water thus again penetrated the western Baltic Sea, and what is now eastern Denmark became an island world. The **Littorina Sea** had – especially in the Gulf of Bothnia – a larger extent than the Baltic Sea today (Fig. 8.10). It was not until

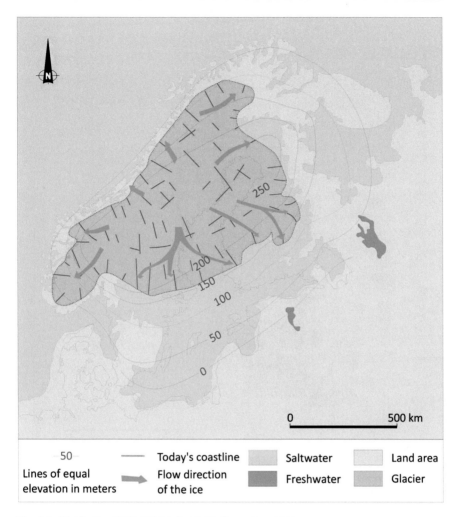

Fig. 8.8   Yoldia Sea (8000–7700 BC) (© Wolfgang Fraedrich)

the isostatic uplift of Scandinavia, which continues to this day, that more and more area was wrested from the Baltic Sea.

As a result of this uplift, the Baltic Sea access to the world sea became narrower, and the exchange of water between the world sea and the tributary sea lost intensity. Due to a large supply of freshwater via the rivers now flowing in from all areas, the salinity decreased, so that a brackish habitat, i.e. a marine ecosystem in the transition between salt water and fresh water, could develop, as documented by findings of the brackish water snail *Limnea ovata* (hence Limnea Sea) and the brackish sand mussel *Mya arenaria* (hence Mya Sea).

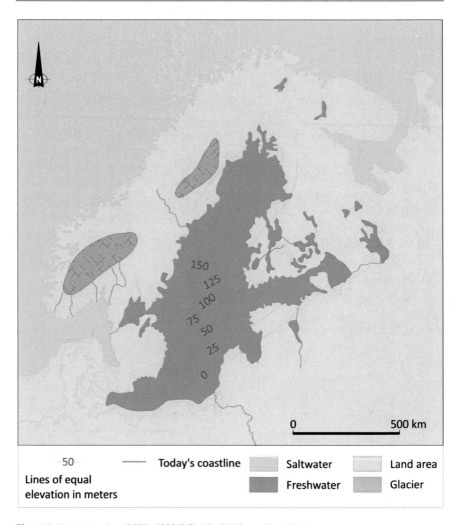

Fig. 8.9 Lake Ancylus (7500–6000 BC) (© Wolfgang Fraedrich)

8.5   Comparison of German Coastal Landscapes

The present coastal landscapes of northern, central and western Europe can only be understood if one considers their Pleistocene and Holocene development. The essential difference between the North Sea and Baltic Sea coastal landscapes in terms of morphological shape and geological structure will not be discussed in detail here. Both coastal areas have glacial deformation in common, but only today's Baltic Sea coastal area was covered by the inland ice during the Vistula cold period. The phase of possible ablation, which lasted about 80,000 years longer, significantly

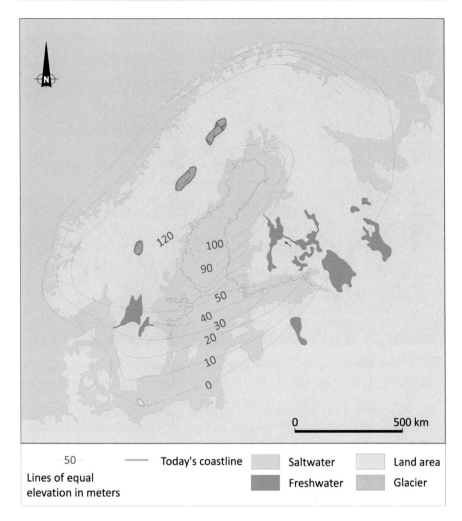

Fig. 8.10 Littorina Sea (6000–2000 BC) (© Wolfgang Fraedrich)

lowered the relief formed during the Saale glacial period. As a result, the relics of the Saale glacial period in the area of the German North Sea coast are limited to a few points. Only on the so-called **Geestkern islands** of Sylt, Föhr and Amrum do boulder clays still protrude above sea level (Fig. 8.11).

The tidal currents (erosion and deposition), which are morphologically important on the North Sea coast with an average tidal range of about 4.50 m, were of no importance in the Baltic Sea area, and in the present day they are of no importance at all due to the only very narrow connection between the two seas.

While in the North Sea coastal area the marshes and the tidal flats were built up by the onset of transgressions as a result of the eustatic sea level rise (cf. Figure 8.5), the flooding of the Weichselian moraine landscapes in the southern Baltic Sea led to a

Fig. 8.11 View from the Geestkern near Wenningstedt (Sylt) into an erosion incision exposing Saale cold period (Warta stage) boulder clay (© Wolfgang Fraedrich)

partly very multi-layered coastline. The Darss peninsula on the coast of Mecklenburg-Vorpommern documents the geologically young development history (Fig. 8.12). At about the beginning of the Littorina Sea, the water level reached the moraine landscape of the Darss region. The Darss was washed around, and the ground moraine landscape became a chain of islands off the coast at that time. The erosive force of the sea eroded the islands and formed, among other things, a striking **cliff** on the north coast of what was then the island of Darss, which today runs right through the island and can still be clearly seen. The sediments eroded from many parts of the moraine landscape in the southern Baltic Sea region were transported in the sea parallel to the coast in many cases and deposited again at other island cores, so that Darss was once again connected to the mainland.

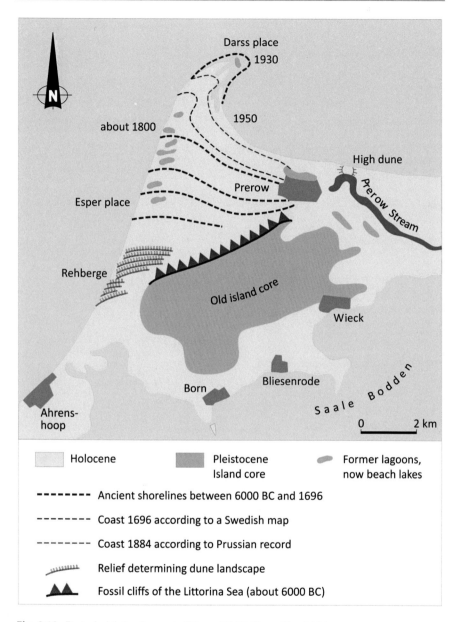

Fig. 8.12 Post-glacial development of Darss (© Wolfgang Fraedrich)

The Next Cold Period Is Coming

9

Wolfgang Fraedrich

Abstract

The final chapter deals with the "burning question" of what the climate will be like in the future. At least in the short and medium term, humanity would like to have concrete indications, because the world's population continues to grow incessantly, and there are global problem areas such as hunger or water shortages.

The current climate development shows a warming trend. Natural and anthropogenic influences play a role, but to date there is no consensus on how significant the influence of humans is for the current climate development. This makes it difficult to arrive at internationally uniform standards for climate protection policy.

Geoscientific research has an important foothold worldwide in climate research, because forecasts for the future lead via the detour of reconstructing the climate of the Earth's historical past.

This chapter should make clear,

- that climate research is complex, time-consuming and costly,
- that the dynamic processes in the atmosphere closely interact with the dynamic processes in the oceans and that, as a consequence, climate can only be understood if these interactions are taken into account,
- what role humans play in influencing the climate,
- that not only humans influence the climate, but that there are also natural factors that trigger climate fluctuations and climate (changes),

W. Fraedrich (✉)
Hamburg, Hamburg, Hamburg, Germany
e-mail: wolfgang.fraedrich@hamburg.de

- how the climate of prehistoric times is reconstructed and what knowledge has been gained so far,
- that it is still problematic or difficult to formulate a truly reliable climate forecast.

Although we humans already know so much today, we cannot yet forecast the climate of the future. The question of "why", which is often asked, is basically quite simple to answer. We still do not know nearly enough. Despite sophisticated technology in the age of the computer, we lack the necessary amount of scientific data to "simulate" a truly meaningful, validated model for the climate of the future by computer.

So we're still in the dark on this one. There are two conflicting predictions in the room:

- Because we humans are turning our Earth into a greenhouse, the atmosphere will heat up. The large freshwater reserves on the Earth's land masses – currently bound up in glacial ice – will melt back and cause eustatic sea-level rise. Shallow coastal areas around the world will be flooded, and in the Pacific whole groups of islands will sink into the sea.
- We are heading for a next cold period, despite all the amplification of the greenhouse effect. Firstly, a rise in world sea levels such as that predicted for the coming centuries – measured against recent developments in the history of the earth – is nothing new, and secondly, it will not be permanent, as there is certainly evidence that the world will also become colder again.

There is not enough space here to go into the details of the scientific discussion of recent years on the climate of the future; it is far too complex for that, and it can be found in numerous other publications. However, it should be made clear here in an overview,

- which points of view are represented, how the individual theories and forecasts are justified or backed up,
- what connection there is between the climate discussion on the one hand and the topic of "ice" – in the broadest sense.

9.1 Global Developments

The climate of the earth is not constant. The controlling parameters are integrated into a very complex structure of effects, comparable to a multitude of different large cogwheels that are all interlocked with each other. If only one of these cogs is turned, the entire system is affected.

Figure 9.1 illustrates the various factors that have shaped terrestrial climate development over the last 600 million years.

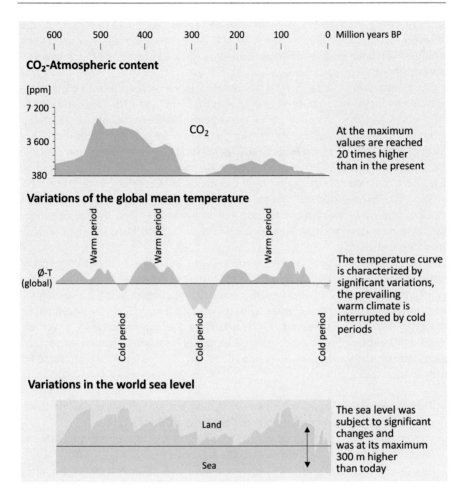

Fig. 9.1 Development of the terrestrial climate over the last 600 million years (© Wolfgang Fraedrich)

The graph shows that today's global mean temperature, which is 15 °C, could rarely have been determined in the past. There were periods when the average temperature was significantly higher, but there were also periods when it was significantly lower. Currently we are living in an epoch in which it is by far cooler than it has been over many millions of years, e.g. in the Mesozoic and the beginning of the New Earth Era. Possibly, however, it is precisely this that has been the prerequisite for mankind to develop into what it is today. We, too, react extremely sensitively to changes – both as biological beings in a complex natural system, but above all because of our civilisation, which is in part highly developed. We are sitting on the moving cogs, so to speak, and we are therefore affected in the broadest sense by all changes that occur in the global system and regardless of where we are on Earth.

Parallels are drawn with the course of the curve for the CO_2 content of the atmosphere and also with the course of sea level – this too was not a constant variable over long periods of time. And it is obvious that species richness and biodiversity on Earth are closely related to climate.

Let us now turn to more recent developments. Level records from all parts of the world show an average sea-level rise of about 0.2 cm/year over the past decades. This results from the warming of the troposphere (the lowest layer of the earth's atmosphere in which weather events take place) and the ocean surfaces down to a depth of about 200 m, which leads to an expansion of the water masses and thus to a rise in sea level. This effect is intensified by the melting of glaciers, which can be detected worldwide, whether in the high mountains, on Greenland or on Antarctica.

The possible consequences of a sea-level rise are classified as very threatening in a time of dense settlement of coastal areas – worldwide. So-called "horror scenarios" have also been developed and discussed for the area of the German Bight with its low-lying marshlands.

But what is the fact now? For the German Bight, the tide gauge measurements of past decades give the following picture: at a roughly constant mean low tide, the mean high tide rose by 0.2–0.3 cm/year, leading to a corresponding increase in the mean tidal range. Such rates of increase have occurred repeatedly in the development history of the Wadden Sea (cf. Chap. 8). They are compensated by a vertical co-growth of the sediment shelf as well as the horizontal, landward shift of the boundaries of the Wadden Sea. However, three preconditions have to be fulfilled for such an adjustment process:

- The rate of increase of the tidal parameters must remain below a critical limit, the value of which is only slightly higher than those mentioned above.
- Furthermore, there must be sufficient sediment volume and sediment mobility to allow for landward relocation and uplift.
- Finally, for landward expansion of marine influence (transgression), a flat, open coastal landscape must be available.

Assuming that the current climate and sea level scenarios occur in the order of magnitude predicted, an accelerating increase in the mean low tidal water of about 0.2 cm/year and in the mean high tidal water of about 0.6 cm/year is to be expected by the year 2100. This leads to an increase of the mean tidal range by 0.4 cm/year. The changed hydrographic situation will be accompanied by meteorological adjustment processes: More frequent storms with higher peak speeds will lead to more frequent severe storm surges. Previously unprecedented storm strengths can then also be expected to produce storm surge heights that are unknown to date.

Anthropogenic interventions in global cycles are considered to be important (co-) causes of this development. These include in particular the increase in carbon dioxide (CO_2) and methane (CH_4) levels in our atmosphere.

9.2   Atmosphere and Climate

Earth is not the only planet in our solar system with an atmosphere, but we do have a "special" atmosphere, because unlike our neighbouring planets Mars and Venus, it is not one gas that is important for the climate-determining radiation budget, but rather a series of trace gases and the strong temperature dependence of the water cycle.

Due to the seasonal differences in solar radiation, heat production on Earth varies greatly from one zonal region to another and, among other things, this triggers the atmospheric control processes, some of which also have a global effect and are variable in their seasonal rhythm (example: the low-pressure foothills of the westerly wind zone that develop on the polar front do not have an effect on the weather in the Mediterranean region during the summer months). Massive interventions in the various ecozones, such as deforestation of the rainforests or anthropogenic overexploitation of the semi-arid tropics, trigger regional climate changes, which in turn also exert an influence on the global system.

The basic principle of our terrestrial radiation budget is that atmospheric gases and water vapor transmit incoming shortwave solar radiation, but absorb outgoing longwave thermal radiation. A reduced water vapour content will trigger a decrease in temperature, as will a reduction in the **carbon dioxide (CO_2) content. In** addition, trace gases such as nitrogen dioxide **(N_2O)**, the trivalent oxygen **ozone (O_3)** and the **methane (CH_4)** already mentioned play a role. In the natural system, these gases and the gas composition of the atmosphere constantly interact with living organisms, so that the Earth's ecosystem is in equilibrium as long as the long-term effective feedback between life and climate (controlled by the radiation budget and the complex composition of the Earth's atmosphere) is given.

Disturbances in this system are regulated as long as this is possible in terms of magnitude and speed. Even the most violent volcanic eruptions, which not only eject solid masses such as lava and ash, but also vast quantities of gases (above all carbon dioxide), have demonstrably not led to any really lasting climate changes, even if there has sometimes been a regionally varying cooling of the mean temperature for a good century, as e.g. after the eruption of the Toba volcano on the now Indonesian island of Sumatra 74,000 years ago. Although they exert a short-term influence on the radiation budget, this is "system-immanent", i.e. remains within the sphere of influence of the self-regulation of the Earth's ecosystem.

The phases of volcanic activity, especially in the geological past, have often been determined using various proxy data. Thus, different CO_2 contents have been reconstructed for the Earth's history. Furthermore, the varying CO_2 content can also be explained by different carbon sinks. Oceans, for example, are able to bind significantly more CO_2 at lower water temperatures. Thus, if it becomes colder on Earth – e.g. due to radiation – more CO_2 is bound in the oceans, which in turn influences the terrestrial climate and intensifies the cooling effect. At the same time, acidification of the oceans occurs, which increasingly deprives carbonate-forming organisms of their basis of life. Changes in marine ecosystems are the result. It is also known that vegetation binds CO_2 and that changes in terrestrial flora – due to both natural changes and anthropogenic influences – at different latitudes thus also control the CO_2 content of the atmosphere. By cutting down the rainforests, for example, humans are thus making a significant contribution to global warming, as more CO_2 is released into the atmosphere.

The atmosphere is therefore part of a very complex system in which it is closely interrelated with the other spheres.

9.3   Anthropogenic Interventions

However, at the latest since the beginning of the industrial age, i.e. since the middle of the 19th century, we humans have been making a decisive contribution to the heating of the earth's atmosphere. Industrial waste gases, exhaust gases from power plants, fuel-powered vehicles and private households, and the methane increasingly produced in landfills and by intensive agricultural production are examples of the fact that carbon dioxide and trace gases are increasingly being accumulated in our atmosphere.

The measurable rise in sea level is an indication of warming, but in a strictly scientific sense this is not yet proof of global warming. A look at the history of the Earth's atmosphere can provide some help here. Trace gases naturally fluctuate strongly, and the same applies in parallel to temperatures (Fig. 9.2).

Carbon dioxide levels were 190 ppm at the peak of the last cold period, 280 ppm before industrialization, and 352 ppm today. Methane concentrations rose from 0.35 ppm to 0.7 ppm over the same period, and then to 1.7 ppm due to human influence. In short, the warmer the Earth has been in its history, the "thicker" the air has been – or vice versa. On the one hand, such a development documents that fluctuations in the composition of the atmosphere can certainly be classified within the framework of natural processes, but it also makes it clear that human interventions must not be neglected.

The current climate discussion is therefore certainly justified. However, it would be wrong to derive a reliable forecast from previous measurements alone. This is precisely what scientists are warning against.

In the following, we will first describe the causes of natural climate fluctuations. Here, very concrete ideas have developed in recent decades. Only on the basis of this can a look into the future be taken, and above all it should be emphasized which current indications are given and how these could be classified in the existing models of thought.

9.4   Natural Climate Fluctuations

Ever since climate changes in the history of the earth could be proven – this was already the case at the beginning of the eighteenth century – research has tried to clarify their causes. And it was already clear early on that the background for changes in the global radiation budget had to be sought. The fluctuations in the composition of the atmosphere already described cannot be held solely responsible

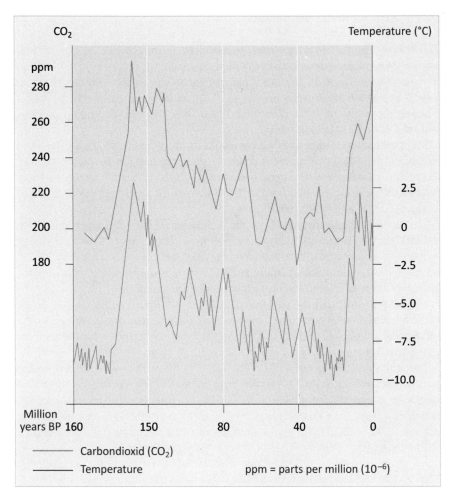

Fig. 9.2 The carbon dioxide content of the Earth's atmosphere runs parallel to the global temperature curve – reconstruction of the climate development of the past 160,000 years via drill core analyses of Antarctic inland ice (© Wolfgang Fraedrich)

for the long-term changes and short-term fluctuations in the global climate. Other components must have played a role.

Already at the beginning of this century theories were developed according to which astronomical influences must play a role, and today it is demonstrably certain: Jupiter is to blame for everything ...

9.4.1 Astronomical Influences

The Earth, as a part of our solar system, is subject to the influence of the gravitational forces of other planets, especially that of Jupiter, which, with its mass of $1.9\cdot10^{27}$ kg, has twice the mass of all the other planets, including all their moons together. These – and to a limited extent also those of Saturn with a mass of $5.68\cdot10^{26}$ kg – influence the orbits of other planets and also their rotation, depending on their position within the solar system (Fig. 9.3).

The **precession** of the earth's axis results from a "spin" of the earth, i.e. it describes a path on an (imaginary) cone – a principle that can be illustrated very well with a spinning top. The period of time in which the Earth's axis completely traverses this orbit covers an average of 21,000 years, although the periods fluctuate between 16,000 and 26,000 years.

The second parameter results from the inclination of the Earth's axis relative to the Earth's orbit, i.e. the so-called ecliptic **obliquity**. This changes within a period of 41,000 years on average with a fluctuation range of just under 2.3° (between 24° 28′ and 22° 0′). At present, the obliquity is 23° 27′, and it is decreasing, which means that the difference between the seasons is becoming smaller.

The ecliptic shift and the precession exert an influence on the climate of the northern hemisphere in particular. This can be explained, among other things, by the fact that the significantly larger part of the land masses extends over the northern hemisphere. Land masses and water surfaces absorb solar radiation differently.

The third parameter is the shape of the Earth's orbit within our solar system, the so-called eccentricity. With a fluctuation range of about 20% (corresponding to 30% fluctuation range of the global solar radiation), the Earth's orbit sometimes resembles an ellipse and sometimes a circle. About every 92,000 years (fluctuating between 72,000 and 103,000 years) it reaches one of the extreme forms, at present it is almost circular.

All three parameters influence the solar radiation on earth and its radiation balance. The different periods result in temporally offset minima and maxima. If

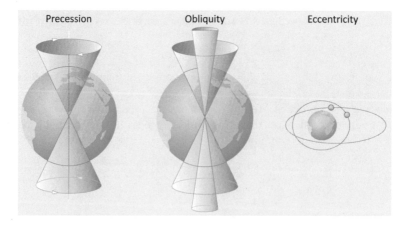

Fig. 9.3 The astronomical influences on the terrestrial climate (© Wolfgang Fraedrich)

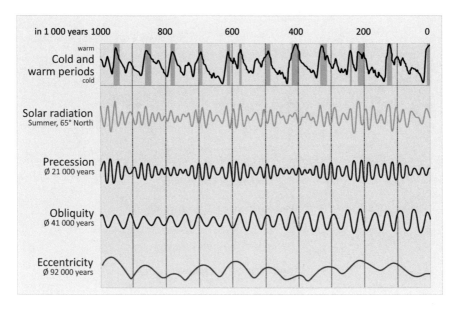

Fig. 9.4 The Milanković cycles (© Wolfgang Fraedrich)

we look at Fig. 9.4, the four different curves show different frequencies and amplitudes. The result is that at certain times, e.g. maxima or minima in the curves quasi double or in the opposite case, e.g. rather neutralize each other. These considerations go back to the Serbian mathematician Milutin Milanković (1879–1958), they are therefore also called **Milanković cycles**.

There are therefore phases in which the climate effects triggered by this can intensify or also cancel each other out. Since it is assumed that the changes in the gravitational field of our solar system will also be very similar in the future, corresponding forecasts can be made for the future course. However, it can be assumed that this will not necessarily be accompanied by a constant alternation of cold and warm periods, because then it would have to be traceable as a continuing process in the history of the Earth, which, however, can be demonstrably refuted today (cf. Chap. 4).

So extraterrestrial influence alone is not enough to keep the globe icing over in certain regions.

9.4.2   Earthly Influences

As early as 1969, the meteorologist and climatologist Hermann Flohn combined numerous aspects into a model in a so-called **multi-factor hypothesis**, which will be briefly outlined below and summarised in Fig. 9.5.

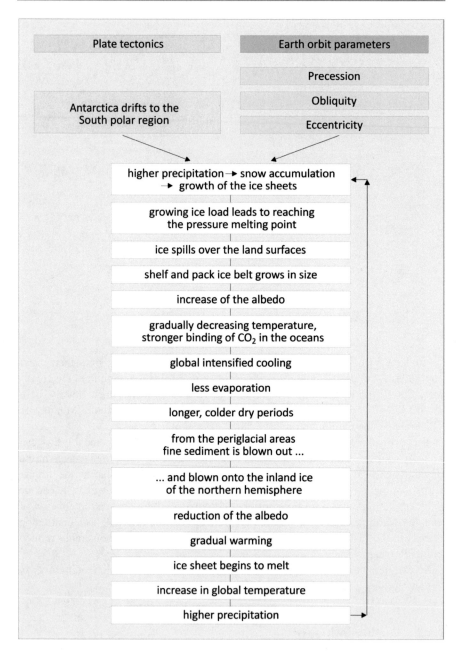

Fig. 9.5 Cold periods and warm periods have several causes – the multi-factor hypothesis (© Wolfgang Fraedrich)

Surprising as it may sound, the alternation of warm and cold periods is also related to plate tectonic processes. In addition, the so-called ice shelf hypothesis plays a role. If one combines the various findings, a model conception emerges according to which the drift of a larger land mass into a position close to the poles is seen as a main cause for the emergence of a cold age. For the present cold age it is the drift of Antarctica into the south polar region, for the Carboniferous/Permian cold ages it was the drift of the then still coherent Gondwana with its (present) subsystems South America, Africa, Madagascar, India, Australia and Antarctica.

With the tectonic movement process, a cycle is currently running – presumably until Antarctica has left the south polar region again.

As early as about 34 million years ago, a steadily increasing winter snow cover began to build up on what is now the continent of Antarctica. Its annual melting triggered an intensifying cooling of the deep waters of the surrounding oceans. Ocean currents extended the cooler deep water to all the oceans, which had no noticeable effect in the lower latitudes but led to the onset of climate deterioration in the higher latitudes at the end of the Tertiary period.

About 15 million years ago, the development of an ice sheet on Antarctica began. The associated lowering of the global temperature finally triggered ice formation in the north polar region from about 3.5 million years before today. A successive lowering of the world sea level was the consequence.

The growth of the Antarctic ice sheet led to an increase in the area of the shelf and pack ice due to its sliding out over the edge of the land mass after reaching the **pressure melting point** at the base, which triggered significant radiative climatic changes. The ice sheets in the northern hemisphere also grew, and the world sea level sank to -145 m in the meantime.

After depressurization, the ice shelf slowly degraded due to a lack of sufficient replenishment, and the cooling of the deep water and the air initiated from the poles resulted in a drop in temperatures even in the tropical region. A reduction of evaporation over the oceans (up to 30%) was the consequence, it became drier worldwide, the ice sheets on the northern and southern hemisphere were no longer sufficiently nourished. The ice sheets of the northern hemisphere were "polluted" at the edges by loess, their surface became darker, thus the **albedo** became lower, during the summer months more radiation could be absorbed and converted into heat, more ice melted with a simultaneous (eustatic) rise of the world sea level.

▶ **Pressure Melting Point** The pressure melting point is the melting temperature due to the load. At an ice thickness of 2500 m, it is $-1.6\ °C$ (Zepp 2014, p. 194). The phase boundary between ice and water is thus below 0 °C at higher pressure. The pressure melting point is thus reached independently of the frictional heat at the base of a glacier by sufficient superimposed load ($=$ glacier ice thickness), the ice deforms plastically, it spreads out to the sides under the pressure of its own weight. This process continues until the load is reduced again to such an extent that the pressure melting point is again undershot.

▶ **Albedo** Albedo (lat. *albere* = to be white) is the ratio of the light striking a non-reflecting surface to the light reflected back. The albedo thus expresses the measure of the solar energy that is reflected back and not converted into heat.

Thus, plate tectonic processes indirectly influence the radiation budget. The partly regionally very different radiation balance initially causes large-scale regional climatic changes, which – via the oceans – change the climate globally.

Further evidence of the influence of plate tectonics on the terrestrial climate is the closure of the Strait of Panama, i.e. the formation of the so-called "Panama Isthmus". This created a land bridge between what is now South America and what is now Central America. The previously unhindered exchange of water masses between the Atlantic and Pacific Oceans was cut off. The Atlantic ocean currents in the mid-Atlantic were now redirected to the north, and today's Gulf Stream system was formed. The resulting significantly higher evaporation rate led to significantly higher precipitation, especially in the area of northeastern Canada, which often fell as snow. This laid the foundation for the cold periods that followed up to the present day.

The processes described, like those of the astronomical components, are subject to periodicity. The build-up of ice sheets, the formation of ice shelves, the increase in albedo and the resulting global cooling are just as periodic as the highly glacial loess dust pollution of the northern hemispheric inland ice masses and its consequences.

If these components are linked together, there should be an average of

- a tundra phase every 5000 years,
- every 22,000 years a significant inland ice spread,
- every 60,000 years.

Man is powerless against these natural changes. They are, so to speak, the astronomical and terrestrial boundary conditions with which we must come to terms.

At the moment, however, we are conducting a large-scale global climatic experiment that has never existed before by nature. We are blowing trace gases into the atmosphere at a speed that the Earth has not experienced since the time of its formation. An experiment with far more than seven billion test subjects!

But there is also something reassuring about looking at climate history: The Earth today is not as sensitive to a change in trace gas content as it was at the end of a cold period. This is because only when there is a lot of ice in the northern hemisphere do the retreating light areas accelerate warming via a positive feedback so massively that sea levels rise rapidly. The melting of ice in the western Antarctic region tends to have a modifying effect in this context because of its smaller area. However, the main ice in the northern hemisphere thawed between 18,000 and 10,000 years ago, causing the level to rise by more than 100 m at that time. The rest is still resting on Greenland today – good for "only" 7 m sea level rise.

Greenland seems to be relatively stable to boot. The island remained mostly icy even during the Eemian warm period 130,000 years ago. At that time, the global temperature was on average 2 °C higher than before industrialisation. However, the trace gas content of the atmosphere is already higher than in the Eemian. So a corresponding but delayed warming is already pre-programmed. And there is no end in sight to emissions. Whether the Greenland ice can cope with this in the long term is part of mankind's "global large-scale experiment".

But where do the findings about the earth's historical processes come from?

9.5   Ice Cores Are Sought-After Climate Archives

Since the 1960s, when planning began for research into climate history, the large ice sheets have become the focus of scientific attention. The way to be able to derive forecasts for future climate development from deciphering the prehistoric climate (paleoclimate) was to be an essential building block in polar research from then on. The finely layered ice would be able to provide information about climate development. For taking ice cores, it was important that the glacier ice was stored as undisturbed as possible; flowing ice cannot give a clear answer due to the constant changes in the ice body, quasi due to a mixing of the ice of different ages.

But how does one extract the information contained in an ice core, since temperature scales are not present there? So-called **proxy data** are used for this purpose.

▶ **Proxy Data** Proxy data are not direct but indirect indicators of climate and thus important climate witnesses. In addition to information from ice cores, pollen, tree rings, corals and sediment layers can also contribute to the reconstruction of the prehistoric climate (= paleoclimate).

When deciphering ice, it is above all the trapped gases. Because when the transition from firn to ice takes place, the gas composition of the prevailing atmosphere is stored. And this gas composition can be analysed. Important parameters are the trapped carbon dioxide (CO_2) and the ratio of heavy to light oxygen. Snow and ice contain two different types of oxygen, firstly the "normal" oxygen and secondly a heavier isotope with two extra neutrons in the atomic nucleus, and these different types of oxygen reflect the temperature change of the air. The higher the air temperature, the higher the concentration of the heavier oxygen in the snow (and therefore in the ice). From the ratio of these two oxygen isotopes, it is possible to infer the temperature conditions at the time when this ice fell as precipitation in the form of snow, even at greater depths of the ice sheet.

The so-called $\delta^{18}O$ (pronounced delta-O-18) is thus a proxy value for the prevailing temperature at the time; it is an indication of the ratio of the stable oxygen isotopes $^{18}O/^{16}O$ relative to the standard ocean water (SMOW = Standard Mean Ocean Water). The temperature can be reconstructed by determining the ratio. Such an analytical procedure is called an isotope survey.

Ice cores have been taken both in the Arctic (on Greenland the GRIP = Greenland Ice Core Project and the GISP = Greenland Ice Sheet Project) and in the Antarctic (among others EPICA = European Project for Ice Coring in Antarctica). The deepest borehole ever drilled was GISP 2 with 3053.44 m of ice and 1.55 m of rock (reached on 1 July 1993). At a depth of 3029 m, some 250,000 years of climate history had been "unlocked" in Greenland shortly before during the period 1989–1992. In contrast, the EPICA project (= European Project for Ice Coring in Antarctica) took the deepest look into the past in the period 1996–2004 through the Dome C borehole (= Dome Concordia). Here, ice up to 900,000 years old has been "opened up" at depth.

9.6   The Greenland Icecore Project: An Ice Core Drilling Brings Surprising Findings

The mighty ice sheets on Greenland and in Antarctica are therefore the climate archives that scientists – mostly international research teams – have focused on in recent decades. Among other things, they came up with the idea of studying the Greenland ice sheet, which is more than 3000 m thick at its maximum.

As early as July 1992, 40 researchers celebrated the first success of the GRIP ("Greenland Icecore Project"), which concerned the technical side of this project: exactly 3029 m of ice had been drilled through, the rocky subsoil had been reached.

The drill core records the climate history of around 250,000 years. Technicians and researchers had to invest 2 years of work, sometimes under the most extreme working conditions. With a diameter of 10 cm and an ice weight of about 30 t, the ice has long been stored in the refrigerated archives of Copenhagen University.

The drill core was obtained from the thickest part of the Greenland ice sheet because, firstly, the amount of precipitation there is relatively high and, secondly, the ice flows there only at a very low speed, so that an above-average number of ice layers were formed under the highest point and have remained in their superimposed position over a long period of time (Fig. 9.6).

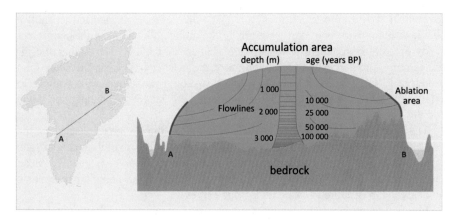

Fig. 9.6 Cross-section through the Greenland ice sheet (© Wolfgang Fraedrich)

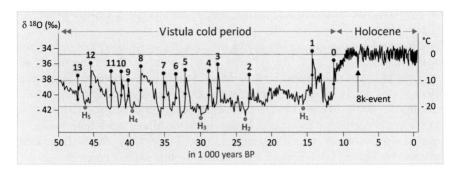

Fig. 9.7 Climate development during the last cold period, Dansgaard-Oeschger events (0, 1, . . .) and Heinrich events (H$_1$, H$_2$, . . .); see also Sect. 9.7 (© Wolfgang Fraedrich)

And these layers of snow and ice have the function of a "geological clock". Since the acid and dust content of the snow changes with the seasons, it is possible to count the individual annual layers precisely – comparable to the growth rings of a tree.

Just 2 years after the drilling was completed, in the summer of 1994, another success was celebrated. The precise exploration of the ice core, which is over three kilometres long, revealed findings that had not been expected after the initial analyses and in this form. The investigations at the GRIP core showed that the average air temperatures had increased or decreased by up to 10 °C within a few decades. These were the so-called **Dansgaard-Oeschger events** (DO events for short, Fig. 9.7).

Of particular interest is the core section that represents the Eemian warm period. The results of the analysis show not only that there were frequent changes in global temperature during the Eemian, but also that in each warmer period the global temperature was significantly higher than today. During the warmer periods, the temperature was on average 5 °C higher. According to model calculations, this meant a world sea level that was up to 8 m higher. While it seems reasonable to link such atmospheric changes to a period of greater volcanic activity during the Eemian warm period, exact evidence is not yet available. The GRIP core also detects a cold period lasting about 5000 years, which probably resulted in an extensive spread of north polar ice.

It is therefore clear that the greenhouse effect is not a recent phenomenon. So can we give the all-clear? No, because we must bear in mind that at that time there were hardly any people who could have been affected by the greenhouse effect. Today, however, the world population (cf. e.g. http://countrymeters.info/de/World) is growing unceasingly ("exponentially"), and no one can estimate what the consequences of a temporary warming of the earth will be for them.

9.7   Atlantic Sediments Show Traces of the Last Cold Period

The Greenland Icecore Project has shown the direction in which glacial geological research must go in the future. The aim is not only to document the climate history of the recent past, but to explore as many details as possible, to fathom the climate history as completely as possible in order to be able to derive forecasts for the future. In addition to studies of cores from glacial ice, especially from the inland ice masses, which allow a very broad retrospective view, analyses of sediments from bodies of water are also becoming increasingly important. Drill cores from lakes on Greenland and marine sediments from the northern Atlantic, together with ice cores, provide the key to the climate history of the past.

Sediment samples from an Atlantic region at about 40° N latitude in the mid-1980s may have provided a further explanation for the advance of Nordic ice (Fig. 9.8, Dreizack borehole, named after the Dreizack seamount off Portugal). Initially, these results were obtained purely by chance: The aim of an investigation in the deep sea of Western Europe was to find out whether the deposition of low-level radioactive material on the Atlantic floor was justifiable. A prerequisite for an assessment of this project was also an analysis of the seafloor sediments. This revealed a low-fossil layer up to 15 cm thick, which gave rise to further investigations. In contrast, the sediment layers below and above were rich in fossils. Further discoveries elsewhere in the North Atlantic revealed comparable results: fossil-poor and, at the same location, the presence of limestone remains from Canada. Partly volcanic ashes from Iceland were also found.

One had soon developed a hypothesis to explain these finds: These materials are typical of sediment deposits on and in glacial remnants – the icebergs that, due to climatic conditions, had the ability to drift far south, much farther than today. Age dating of the layers showed that their deposition must have taken place in the high glacial period, about 14,000–21,000 years ago in several phases of about 1000–2000 years in length. This would correspond to the respective inland ice maxima of the high glacial period, in the course of which the main end moraines of the Brandenburg, Frankfurt and Pomeranian stages, which have been described several times, were built up.

It is possible that this relatively short-term fluctuation of the climate cannot be explained exclusively by the multi-factor hypothesis published in 1969. It must be assumed that the complex of causes must be extended by a further link, which has not yet been taken into account in the multi-factor model. This is the so-called "**Heinrich events**" (Heinrich after the German geologist Hartmut Heinrich, who first discovered the sediments in 1983, and event = event). This refers to the temporary slipping of inland ice masses into the North Atlantic for several centuries as a result of a relatively rapid increase in the mass of these ice sheets. The consequence of these 'events' was an increased supply of fresh water with the consequence that no more deep water was formed in the North Atlantic with which the global ocean current system (Fig. 9.8) could be kept going in these latitudes. The cold and specifically heavier deep water forms "normally" in the European North Sea where the water of the North Atlantic Current gradually cools and begins to sink.

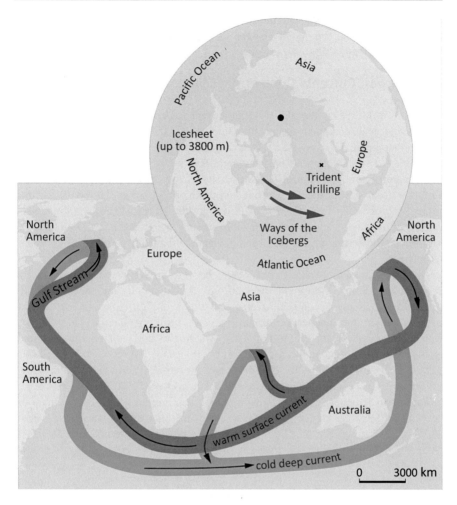

Fig. 9.8 The global ocean current system and the tracks of Heinrich events – simplified (© Wolfgang Fraedrich)

On its way back into the open Atlantic, the deep water overcomes the ocean ridge between Greenland and the British Isles. In the process, the water sinks more than 1000 m into the depths, similar to a waterfall, and mixes with the adjacent water layers due to the turbulence triggered.

A drop in salinity of only 0.2% is already sufficient to prevent the formation of deep water. The consequence was therefore that the Gulf Stream, Europe's "central heating system of the present", did not flow as far into the North Atlantic. On the one hand, this allowed icebergs to drift far to the south, but it also led to significant cooling in northern and central Europe over a relatively short period of time. This explains each new ice advance within the Weichselian cold period. When the equilibrium in the mass balance of the ice sheets was restored, the ice did not slide out in such large quantities. The salinity in the North Atlantic rose slightly again, the

Gulf Stream shifted its reversal point further north again, and thus it became warmer again for a short time, the inland ice even melted back until it could advance again.

9.8   Ocean Currents Control the Climate

Thus, if an influence on the global ocean current system has a climate controlling effect, then it is reasonable to assume that the oceans as a whole play a major role in the Earth's climate.

In order to supplement the previous remarks, the ocean currents are to be the focus of consideration. Figure 9.8 clearly shows the course of the global ocean current system. With respect to Europe and the North Atlantic, the fact that warm surface water coming from the tropics flows into the North Atlantic via the Caribbean (Gulf of Mexico, hence the name Gulf Stream) is significant. Its foothills even flow around the northern tip of the Scandinavian mainland and penetrate as far as the Barents Sea.

This far northward advance only succeeds because the surface water is pulled by an invisible undertow. Between Greenland, Spitsbergen and Iceland there is a "hole in the ocean". Every second, an average of half a million cubic metres of cold seawater sinks into deeper ocean layers. The resulting suction pulls in the approaching tropical surface water. This "water elevator", called thermohaline circulation, is part of an enormous flywheel in the oceans. The cooled water masses sink 3000–4000 m deep to the ocean floor and flow back south at a speed of a few centimetres per second in the depths of the Atlantic Ocean.

The sinking of ocean water at such high latitudes is related to the salt content of the Atlantic. Seawater contains 34.9 g of salt per litre and therefore only freezes at − 1.8 °C. Once the water in the North Atlantic has cooled to this temperature, the first ice crystals begin to form. But fractions of a degree before the freezing point, the water becomes so heavy that it can penetrate the underlying water layers and sink.

However, even the slightest decrease in salt concentration blocks the circulation. Even before the cooling water would have the necessary gravity to sink, it would freeze to ice. Under closed ice layers, however, water circulation no longer takes place. The consequence would be that the outflow for the Norwegian Current, an offshoot of the Gulf Stream system that flows into the Arctic Ocean, would be blocked and the ocean current would come to a standstill here. Its reversal point would shift to warmer latitudes.

This makes it clear: the climate of the northern latitudes is determined by the Gulf Stream, but the Gulf Stream is also dependent on the climate of the northern latitudes. This is another example of how complex our terrestrial climate system is.

9.9   Can There Be a Convincing Climate Forecast?

Figure 9.7 shows quite clearly that the climate has been more constant over the last 10,000 years or so than it has been for a long time. And yet we are (justifiably) concerned about the climate of the future.

Forecasts of future climate, e.g. with regard to the global average temperature or the mean precipitation of an entire climate zone over one or more decades, are associated with the following uncertainties:

- External factors influencing climate.
  On the one hand, these can be of natural origin (e.g. greatly changed solar activity, increase in volcanic activity), but on the other hand it is the uncertainties associated with human influence. Neither population trends nor changes in consumer behaviour, energy consumption, the use of existing energy sources, technological development and agricultural production can be precisely deter- mined. This, for example, has prompted the International Panel of Climate Change (IPCC) to develop various emission scenarios for greenhouse gases, because this is the only way to take account of the different development possibilities of global society.
- Limited knowledge of the climate system.
  The results of research into the climate system and its dynamics available to date are still limited, e.g. with regard to the chemical and physical processes in the atmosphere, which influence cloud formation, among other things. Numerous questions also remain unanswered regarding the influence of cloud formation on the atmospheric radiation budget. Knowledge of the interaction between the atmosphere and the oceans is also still incomplete.
- Shortcomings of climate models.
  Forecasts of the climate of the future are based on computer model simulations that attempt to represent the complexity of the climate system with its dynamics and the influencing variables of external factors. The quality of these simulations depends on the performance of the computers and also on the available database. Currently, the spatial resolution of global atmospheric models is a grid of more than 100 km width. Small-scale processes (e.g. cloud formation) are not yet covered by this.

This raises the central question for the future as to what influence the further increase in anthropogenic greenhouse gas emissions will have on the climate of the twenty- first century and beyond. And if one assumes that the climate will change, the question of the consequences of the expected climate change also remains open.

Taking into account all the uncertainties mentioned above, numerous scenarios of possible climate states at the end of the twenty-first century and beyond are possible (Fig. 9.9).

According to the current state of science, the approaching "climate catastrophe" widely discussed in the media cannot yet be predicted with certainty.

Arguing for a much warmer climate in the next century:

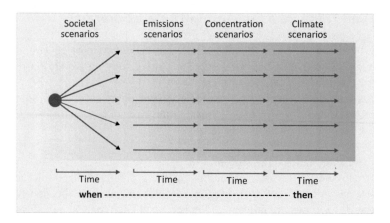

Fig. 9.9 Model representation of the "colourful variety" of possible climate scenarios (© Wolfgang Fraedrich)

- Greenhouse gas emissions are growing worldwide.
- If we consider that greenhouse gases have a long residence time (in some cases up to 200 years) and a long-term effect in the atmosphere, we cannot even feel the consequences of all the trace gases released into the atmosphere today.
- All efforts to find a political solution to the greenhouse gas problem have so far been almost fruitless. Mere declarations of intent, as repeatedly agreed at various climate summits (cf. e.g. https://de.wikipedia.org/wiki/UN-Klimakonferenz), hold little promise for limiting emissions. After all, economic growth still has priority.
- The world population is growing at a rapid pace, by several people every second (cf. e.g. http://countrymeters.info/de/World). They all want to be provided for, which demands further "concessions" to the environment.
- Presumably, we are living in a "super warm period" due to increasing anthropogenic climate forcing.

The far-reaching effectiveness of anthropogenic influences must be countered:

- Climate fluctuations are a demonstrable part of the Earth's ecosystem. The Earth has experienced both significantly warmer and colder periods in the past. In particular, GRIP analyses have shown that the last warm period, the Eemian, was significantly warmer than the Holocene, even at its so-called climate optimum a few 1000 years ago.
- The currently very slow sea-level rise is – at least so far – far from reaching the dimensions of the early Holocene. Since increased melting of the ice sheet on Greenland and the islands of the Nordic Seas will also lead to increased precipitation again and thus to a greater supply of snow in the nutrient areas of the glacier regions, there can probably hardly be a worrying increase in annual sea-level rise

of more than about 0.2 cm/year. In this context, it must be emphasized that a melting of the Arctic drift ice will have no effect on sea level ("ice cube effect").

- For the Antarctic region, no significant climate impact is currently expected due to the average surface temperatures of −20 °C in summer.
- Also in the Holocene there was a decrease in temperature of up to 2 °C for shorter periods of time, such developments could compensate for anthropogenic warming.
- According to the findings on the changes in the Gulf Stream system, a temporary global warming could result in an exactly opposite development for northern and central Europe. But even for this – if it should really be so (so far it is only a hypothesis!) – one would have to apply a period of at least 1000 years.

A short- or medium-term, but above all reliable climate forecast is therefore not possible in either direction. In the long term, however, it is clear that the next cold period is bound to come! According to the analysis of the Antarctic ice core "Dome C" and the information provided by the scientists of the Alfred Wegener Institute in Bremerhaven involved in it, it will in all probability take another 15,000 years until the next ice age. But even a cooler phase of the dimensions of the "Little Ice Age" in the nineteenth century would lead to a catastrophe in densely populated Europe, including the high altitudes of the Alpine region that are now being cultivated.

Glossary

Ablation debris Material that glaciers remove through their erosive action and leave lying elsewhere as they melt is called ablation debris.

Ablation Erosion.

Ablation Melting and evaporation on glaciers, firn and snow surfaces, mostly at the surface.

Accumulation Another name for deposition, means settling of transported rock material, sedimentation.

Active layer A technical term used in the Anglo-Saxon world for the topsoil layer in periglacial areas that thaws briefly in summer.

Adjustment measurement A geological fieldwork method that aims to determine the position of the longitudinal axis of rock bodies in unconsolidated sediments (e.g. in flowing soils). The longitudinal axis adjustment measurement allows conclusions to be drawn about the predominant direction of movement of the transported sediments.

Aeolian Caused by the wind (after Ailos, the Greek god of the wind).

Albedo A measure of the ratio of light incident on a non-reflective surface to the light reflected back.

Altitude gradation Especially the high mountains of the earth show a climatically conditioned altitude gradation due to the decreasing temperatures with increasing altitude, which can be recognized by the respective vegetation. High mountains basically show a nival stage, a stage of perpetual snow and ice. In the course of the current global warming, the altitude boundaries or transitions of the individual altitude levels are shifting. For example, it is predicted that the ice cap of the 5895 m high Mount Kilimanjaro in Tanzania will completely disappear between 2022 and 2033.

Bedload A term for materials that have been removed by glaciers, transported and redeposited elsewhere – often hundreds of kilometres away.

Börde Name for the lowland bays reaching into the low mountain zone of Europe from the North German Lowlands, such as the Jülich or Magdeburg Börde. The geology of the Börde is characterised by loess deposits, some of which are more than 1 m thick.

Boulder clay A term for moraine deposits of the latest cold period (Weichselian/ Würmian cold period) and thus of the so-called young moraine landscapes.

Unlike the boulder clay of the old moraine landscapes, boulder clay has a limestone content.

Boulder clay Moraine deposits of the Saale−/Riss- and the Elster−/Mindel-cold time, they are located in the old moraine landscape. Due to their age, the lime residues they contain have already been dissolved out by chemical weathering.

Brandenburg main ice edge Main ice edge location.

Build-up layer Active layer.

C14 dating A method of absolute age determination in which the age of sediments can be determined by determining the radioactively decaying carbon C14, whose half-life is 5730 years. In the most extreme case, C14 dating can be used to date sediments with an age of up to 70,000 years.

Carbonic acid weathering An example of chemical weathering. Easily soluble calcareous rock (limestone $=$ $CaCO_3$/calcium carbonate) reacts with natural carbonic acid resulting from the chemical combination of water (H_2O) and atmospheric carbon dioxide (CO_2). The reaction equations for the dissolution process are $H_2O + CO_2 \rightarrow H_2CO_3$ and $H_2CO_3 + CaCO_3 \rightarrow Ca^{2+} + 2\ HCO_3^{2-} -$ so the calcium carbonate decomposes to calcium ions and hydrogen carbonate by the action of the carbonic acid.

Channel lake Elongated lake formed and preserved in a subglacially formed channel system.

Channel Valley Gutter System.

Clay mineral End product of weathering ($=$ new weathering formation). The clay minerals are the nutrient stores of the soils, in other words: soils without clay minerals are infertile.

Cliff Another term for cliff.

Climate (change) Long-term climate change, e.g. the change from a cold period to a warm period.

Climate fluctuation Short-term, not too severe change in climate (e.g. a colder phase within a warm period, cf. Little Ice Age and stage).

Climate history Documentation of the development, fluctuations and effects of the earth's climate. It can, for example, cover geological periods over several million years or only a few decades in the recent geological past.

Coalification The process of transformation of vegetable matter into coal, which is characterized by the so-called incarbonation series: wood – peat – lignite – hard coal – anthracite – graphite.

Cold period Period of Earth history within a cold age, i.e. an epoch in which there is ice on Earth, during which the average global temperature is significantly below that of the respective warm periods. During the Weichselian cold period (approx. 75,000–10,000 years before today), the average global temperature was up to 4 °C below today's average of 15 °C.

Concordance Term for an inherently parallel superposition of rock strata without disturbance.

Corrasion The natural process of mechanical working of rocks, e.g. by other rocks. Sandstorms also cause corrasion on rocks.

Corrosion (lat. *corrodere* = to decompose, to eat away), oxidation of metals.

Crevasse A break or crack in glacial ice due to increasing compressive or tensile stress in the ice. The ice is not plastic enough to withstand the stresses. Crevasses form primarily at the edge of the ice and where ice flows over terrain irregularities.

Cryogenic Caused by ice, derived from Greek (Gr. κρυος = frost, ice).

Cryoturbation Periglacial forming process in the topsoil as a result of thawing and refreezing of the active layer.

Dansgaard-Oeschger event Short-term climate fluctuation in the Weichselian cold period named after the Danish paleoclimatologist Willi Dansgaard and the Swiss physicist and climate researcher Hans Oeschger.

Dead ice Glacial ice that remains at the edge as the ice melts back and is no longer incorporated into the active, flowing ice.

Deep erosion Erosive force acting in the depths. Deep erosion in flowing water leaves deep incisions in the landscape, a process that is more intense when it occurs in hard, resistant rock. Notched valleys are a typical result of the prevalence of deep erosion.

Dehydration A process effective in the periglacial zone that refers to the withdrawal of thaw water from adjacent, still frozen rock layers along the dew front.

Density The density of a material expresses the ratio of weight to volume (g/cm^3 or kg/m^3).

Deposition Sedimentation.

Detersion A type of glacial relief formation, i.e. caused by glacial ice, in which the solid rock bedrock is mechanically stressed by the material carried at the base of the glacier. Glacial scars, for example, are a result of detersion.

Detraction The breaking out of the solid rocks at the base of the glacier by glacial processes.

DO event Dansgaard-Oeschger event.

Drainage area The area of the glacier in which the glacier loses mass mainly through melting processes.

Drill core Rock or ice sample that can be extracted by drilling to greater depths. Drill cores are usually divided into segments of 1 m (this makes them technically easier to obtain) and have a diameter of about 10 cm. The analysis of the cores gives an insight into the stratigraphy of the drilled area.

Drumlin Glacial fill form. Drumlins are elongated narrow ridges that often occur in the company of other drumlins (drumlin fields). They consist of moraine material and are unstratified in structure.

Dune Aeolian form of deposition. Dunes are predominantly composed of fine sand that is eroded by the wind and redeposited elsewhere. Dune formation was common in periglacial areas during the cold periods.

Earth's New Era The most recent period of Earth's history (beginning 66 million years ago), also known as the Cenozoic, which is divided into the Paleogene (66–23 million years ago), the Neogene (23–2.588 million years ago), and the Quaternary (since 2.588 million years ago).

Eccentricity Change in the shape of the earth's orbit, which at times resembles a stronger ellipse, at times more like a circle.

Ecliptic Inclination of the earth's axis with respect to the earth's orbit.

Eisrandlage An ice margin marks a phase of glacial standstill that has existed for several decades, usually by means of a prominent terminal moraine. During the last cold periods, new ice margin layers were formed again and again due to the constantly oscillating glacier margin.

Erosion Another term for erosion. Erosion acts both to the side (= lateral erosion) and in depth (= deep erosion).

Erratic block Erratic blocks are very large boulders, often several metres in diameter, which are sometimes moved over long distances by glaciers. The erratic blocks in northern Germany are rocks (e.g. granites, gneisses) from northern Europe that came from Scandinavia as a result of the Nordic inland ice of the last three cold periods.

Eustasy Intrinsic fluctuation of sea level, e.g. due to melting of inland/glacier ice on the mainland.

Evaporite A type of sediment formed by precipitation of minerals from water due to high evaporation rates under arid climates (= desert climates).

Exaration A process of glacial dynamics, it describes the excavation, scaling and folding of pre-existing loose and solid rocks at the glacier front.

Firn (Old High German *firni* = "previous year") is so-called old snow, which is at least 1 year old and melts into larger firn grains through frequent thawing and freezing. Firn has a density of 0.4–0.8 g/cm^3.

Floating balance Isostasy.

Flowing soil Slow-flowing unconsolidated sediment that is characterised by extensive water saturation of the soil and slides downwards in accordance with the slope gradient – due to gravitational forces. The flowing process is called soil flow (solifluction).

Fluvial Fluviatil.

Fluviatile Caused by flowing water.

Fluvioglacial Caused by meltwater.

Forces, exogenous A distinction is made between forces and processes, i.e. the forces that trigger processes and the processes themselves. An example: The force is the glacial ice, the process is the glacial erosion/abrasion controlled by the flow of the ice.

Frankfurt main ice edge location Main ice edge location.

Frost blasting Dominant mechanical weathering process in subarctic climates (and thus periglacial regions). Meltwater penetrates rock crevices and fissures, freezes during the winter, increases its volume and thus mechanically stresses the rock. Frost blasting is particularly effective when there is a frequent and, above all, brief temperature change around the freezing point.

Frost cracking A term commonly used in the English-speaking world for the formation of contraction cracks on the soil surface due to frost action.

Geest Landscape designation for the regions of northern Germany shaped by glaciers and their meltwaters. According to their age, they are divided into low Geest (influenced by the Saale glacial period) and high Geest (influenced by the Weichsel glacial period).

Geestkerninsel An island that has a core of cold-age sediments rising above sea level, but to which other (younger) sediments have been added. These younger sediments may be, for example, marsh or beach wall formations.

Glacial scars Traces in the solid bedrock formed in the direction of flow of the glacier ice by detersion.

Glacial series Term introduced at the beginning of this century by Albrecht Penck to describe the form society in the glacial margin area. Ground moraine landscape with tongue basin and tongue basin lake – main end moraine – outwash plain – glacial valley.

Glacial Caused by glacial ice.

Glacier cirque Cirque.

Glacier fluctuations Changes in the mass balance of a glacier that are expressed by the ice either gaining mass and thus advancing or losing mass and thus receding.

Glacier gate An "opening" in the area of the glacier margin – usually in valley glaciers – from which meltwater escapes.

Glacier tongue When glacier ice flows downwards in valleys, they are narrow and elongated. Such ice bodies running towards the ice edge are called valley glaciers.

Gravel plain Areas in the foothills of the Alps that can be equated with the outwash plains in north-central Europe in terms of their formation, but which, in contrast, are built up from coarser, but also stratified unconsolidated material.

Gravitational Controlled by gravity.

Greenhouse effect Similar to a greenhouse – there the short-wave solar radiation penetrates the glass, but the long-wave heat radiation is retained under the glass – the short-wave solar radiation reaches the earth's surface, but the long-wave heat radiation is retained by carbon dioxide and trace gases. The increase in the proportion of certain trace gases and carbon dioxide in the atmosphere leads to global warming.

Ground moraine Glacial form of deposition that only emerges after the ice has melted. These are subglacial (formed under the ice) deposits. A distinction is made between flat, flat-wavy and dome-shaped ground moraine landscapes. While the flat ground moraines are formed in the area of high ice thickness, the dome-shaped ground moraines are typical for the crevasse-rich ice margin area.

Gutter system Gully systems are formed by the erosion of meltwater flows beneath the ice. They are therefore subglacial and at the same time fluvioglacial forms of ablation.

High glacial period Period of the Weichselian cold period from about 21,000 to 14,000 years before present. During the high glacial period, for example, the Nordic ice sheet reached its maximum extent.

Humps When glaciers flow over rocky bedrock, the grinding action of the ice at its base usually creates streamlined rock humps.

Hydrolysis Literally means "dissolution of water". Hydrolysis is one of the most common chemical weathering processes in which water (H_2O) must be involved. Especially siliceous rocks (i.e. magmatic and metamorphic rocks composed to a large extent of silicon dioxide $= SiO_2$) are weathered by hydrolysis. During this process the H_2O dissociates to H^+ and OH^-. The H^+ ions attach themselves to the crystal lattice of the minerals of the rock and cations are exchanged. This leads to the expansion and finally to the disintegration of the original crystal lattice.

Ice climate The ice climate characterizes the climate zone of perpetual frost. According to the climate classification established in 1961 by W. Köppen and R. Geiger it is the so-called "E-climate" (warmest month <10 °C), in the climate classification established in 1963 by C. Troll and K.-H. Paffen, it is the so-called "I1-climate" (high polar ice climates), according to the current "effective climate classification" according to A. Siegmund and P. Frankenberg from 2014 it is the "Fh-climates Fh2, Fh3, Fh4" (annual mean temperature < -10 °C).

Ice Sheet Ice Sheet.

Ice wedge A periglacially formed form caused by frequent frost change in subarctic regions. Ice wedges can form recently, but are also detectable as fossil ice wedges throughout Central Europe. Depending on their formation, a distinction is made between epigenetic ice wedges (formed after rock formation) and syngenetic ice wedges (formed with rock formation). Syngenetic ice wedges are therefore only found in unconsolidated sediments.

Inglacial Intraglacial.

Inland ice Term for an extensive glacier region which, as e.g. on Greenland, is distributed over a large area of land masses (>50,000 km^2) and whose ice thickness is usually >1000 m.

Intraglacial Within the glacial ice.

Isostasy Floating equilibrium of the Earth's crust on the viscoplastic upper mantle.

Isothermal Line of equal temperature. Isolines are lines of equal valence.

Kame Kames are fluvioglacial sedimentary forms, namely meltwater sediments in the marginal area of a glacier; mostly sedimented in glacier margin crevasses. They are essentially composed of sand. Their structure is characterized by stratification.

Kaolin sand A sand that is formed in warm-humid climates. It is very light in colour. The kaolin sands of Central Europe are relicts of the Tertiary period, in which corresponding climatic conditions were typical. On the North Sea island of Sylt, for example, kaolin sands are found. Kaolin consists mainly of the clay mineral kaolinite, which is used as a raw material for porcelain. It has a very high aluminium content and is formed by intensive chemical weathering of rocks rich in feldspar, in which soluble substances are leached out and insoluble substances are enriched. Kaolin itself is not sand, but a loose rock that also contains quartz.

Kar Glacial erosion form in the upper part of the former glacier feeding area. Originally a firn depression that was deepened in the course of glacial erosion.

Karst Is a collective term for all terrain forms that occur where permeable, water-soluble rock (e.g. limestone, gypsum, salts) is present. Karst forms occur both on the surface (e.g. tropical cone karst) and underground (e.g. caves) as a result of leaching/solution weathering, in which surface and groundwater is involved.

Late glacial The youngest and last period of the Weichselian cold period from about 14,000 to 10,000 years before present. During the Late Glacial, the Earth gradually warmed, the large inland ice masses increasingly melted back and the world sea level rose again significantly (at an average of 1 cm per year).

Lateral erosion Erosive force that acts to the side. If rivers or meltwater streams flow through e.g. flat landscapes in loose rock, pronounced lateral erosion takes place. The formation of river loops (meanders) is the result.

Lateral moraine Rock material of varying grain size composition transported and deposited at the edge of a downhill flowing valley glacier.

Lignite seam Term for a lignite-bearing layer within a rock formation.

Little Ice Age Term for a cold period in the late Holocene with a peak around the middle of the nineteenth century.

Little Ice Age Little Ice Age.

Loess Aeolian fine sediment whose present distribution is due to the processes during the Weichselian. The fine materials blown out of the periglacial areas with little vegetation at that time were partly transported over hundreds of kilometres by the wind and were only deposited where natural obstacles stood in the way of further transport (northern edge of the low mountain range zone) or where wind conditions changed, e.g. in the basin locations of the low mountain range zone.

Main ice margin Distinctive wall-like and elongated elevation (end moraine), more than 100 km long, which marks a colder phase within a cold period and its ice margin. For the most recent cold period, the Weichselian, these are the Brandenburg main ice margin (approx. 18,000 years ago), the Frankfurt main ice margin (approx. 15,000 years ago) and the Pomeranian main ice margin (approx. 14,000 years ago).

Marsh Landscape designation for the flat regions along the German North Sea coast that are barely above sea level. The marshes are geologically built up from marine and perimarine sediments and they were only formed in the Holocene due to the eustatic sea level rise and under the influence of the twice-daily changing tidal currents.

Mass balance The mass balance of a glacier results from the ratio of ice loss, predominantly in the glacier's glacier region, and ice gain in the glacier's nutrient region. If the ice loss predominates, the mass balance is negative and the glacier margin retreats.

Medial moraine A medial moraine is formed when two valley glaciers flow together from different valleys and the lateral moraines, which are aligned with each other, unite to form a moraine that runs along the middle of the valley glacier.

Metamorphosis In glaciology, the change from snow to firn to ice.

Moraine A collective term for glacial forms of deposition. The rock material that a glacier removes and transports is deposited either under the ice (dome-shaped and flat ground moraine), at the front of the glacier (terminal moraine: sentence terminal moraine, upsetting moraine) or – in the case of mountain glaciers – at the sides (lateral moraine). If two valley glaciers flow together, the lateral moraines develop into a medial moraine. Moraine material is unsorted and thus also unstratified.

Morphodynamic cycle Designation for the dynamic system, with which the relief-forming exogenous (external) forces are summarized.

Nivation The process by which relief is shaped by the direct action of snow on bedrock, as well as by movement, pressure, and (snow) meltwater.

Notched valley Valley shape created by the erosive force of flowing water with a V-shaped valley cross-section (hence also V-valley). Notch valleys are typical of mountain landscapes. Deep erosion predominates over lateral erosion.

Nourishing area The area of a glacier where the glacier receives ice supply via sufficient fresh snow.

Old moraine landscape In north-central Europe as well as in the foothills of the Alps, this is the common term for those areas that were formed during the Saale/ Riss and Elster/Mindel glaciations and were no longer formed by the glacial ice of the Weichsel/Würm glaciation.

Os A fluvioglacial form of deposition. Oser are formed subglacially or inglacially in tunnels or gullies created by meltwater. Oser build up predominantly of sand and gravel and are stratified in structure.

Outcrop An outcrop is a section of the geological subsoil that provides an insight into the rock structure and the bedding conditions. Outcrops can be created by natural processes (e.g. a cliff), but in many cases they are the result of human intervention in the natural environment (e.g. a sand, gravel or crushed stone pit, a quarry, a construction pit or a road cut).

Oxidation Reaction of a substance with oxygen (O_2). Example: Iron reacts with oxygen dissolved in water: $2\ Fe + O_2 \rightarrow 2\ FeO$ to iron(II) oxide.

Paleoclimate Prehistoric climate.

Periglacial area At the edge of the ice.

Periglacial At the edge of the ice.

Permafrost (ground) Permafrost.

Permafrost (ground) Another term for permafrost (ground). Permafrost is defined as soil and sediment/rock that is permanently frozen for at least 2 years. Permafrost soils have different thicknesses depending on latitude and elevation in mountains.

Plate tectonics Comprises all endogenous activities (processes) in the course of which the lithospheric plates, from which the outer shell of the Earth is built up, move relative to each other. The velocities of horizontal plate movements reach a maximum of 13 cm/year.

Plateau glaciers An extensive glaciation of land masses with an area of $<50,000\ km^2$.

Pollen analysis With the help of pollen analysis it is possible to reconstruct the flora of prehistoric times. The pollen of fossil plant remains is compared with that of present-day plants and conclusions can thus also be drawn about the prehistoric climate.

Pollen Pollen.

Pomeranian main ice margin Main ice edge layer.

Prehistoric climate Climate of the geological past (= paleoclimate). The prehistoric climate – especially that of the past cold age – is being researched in great detail today, because it is hoped that it will also provide information about the future climate.

Pre-Quaternary Temporally before the Quaternary, i.e. older than about 2.6 million years, and lithostratigraphically below the Quaternary.

Principle of actualism The geologist makes observations of geological forces and processes in the present (so-called recent mechanisms) and, in explaining geological processes, assumes that they also took place in the same way in prehistoric times.

Processes, exogenous and endogenous Forces, exogenous and endogenous.

Receiving water A surface water body that receives water from other water bodies and drainage systems. The term is derived from the phrase "before the flood".

Regression Retreat of the sea (e.g. as a result of eustatic sea-level fluctuations).

Regulation Melting of ice when pressure increases (freezing point decrease) and refreezing when pressure decreases (freezing point increase).

Rock Preparation Weathering.

Sander Extensive outwash plains formed in front of the terminal moraines of the inland glaciations. Meltwater transported finer and coarser material out of the glaciated area and sedimented it according to the prevailing flow velocity and thus sorted by size in the glacier forefield. Sander are therefore fluvioglacial forms of sedimentation.

Sedimentation (lat. sedimentum = sediment), deposition of particles of various sizes that are or have been transported by wind, flowing water, ocean currents or glacial ice. The size of individual sediment particles depends on the transport energy that is or was the basis of the system. The higher the energy, the heavier the transported particles.

Séracs Towering formations of glacial ice that occur where glaciers suddenly have to overcome a steep gradient. The ice is often stretched in different directions, so that these towering pieces of glacier are often criss-crossed by deep longitudinal and transverse crevasses.

Set end moraine Terminal moraine.

Shelf base The area already below sea level that can still be assigned to the continent (continental margin). It is delimited towards the land by the coastlines, towards the open ocean by the so-called continental slope. Shelf seas are shallow seas such as the North Sea.

Situmetry Settling measurement.

Snow line Lower limit of the area where snow does not completely disappear even in summer. Above the snow line, firn and later ice forms.

Solifluction Soil flow, a process in periglacial areas that takes place in the so-called thawing layers (active layer).

Soll (= waterhole) A small depression, usually only a few square metres in size, formed after the final thawing of dead ice remains. Sinkholes are only found in the young moraine landscape. A soll is a hollow form that has been excavated by ice, but has only been revealed by the melting of the ice. In principle, it is a "negative depositional form".

Solution weathering A process of rock preparation by chemical processes, e.g. carbonic acid weathering (dissolution of calcareous rocks) or hydrolysis (dissolution of silicate rocks such as igneous rocks).

Specific gravity Density.

Spores Plant seeds.

Stage Period of time within a cold period that is characterized by an advance of the glaciers with a longer standstill phase (climatic oscillation).

Stone ring Structural soil produced by frost heave processes in periglacial areas.

Stratigraphy A branch of geological science that aims to classify rocks according to their chronological order of formation, taking into account their inorganic and organic characteristics and contents, and thus to establish a time scale for dating geological processes.

Structural soil Periglacial (cryogenic) patterns on the soil surface (stone ring).

Subglacial channel Gully system.

Subglacial Under the ice.

Terminal moraine Form of deposition created by glacial ice (i.e. glacial) in regions where glaciers shape the relief. A distinction is made between set end moraine and compressive end moraine. Set end moraines are always formed when the ice margin remains constant (with a balanced mass balance of the glacier) and are usually the relic of a one-time glacier advance, while set end moraines are typical of a so-called oscillating ice margin, in which new ice advances have occurred repeatedly due to short-term climatic fluctuations.

Tongue Basin Glacial erosion form in which essentially the exaration has been effective. Many of the tongue basins formed during the Weichselian/Würm glacial period contain lakes today, so-called tongue basin lakes.

Trace gas A gas that occurs in the Earth's atmosphere in very low concentrations.

Transgression Advance of the sea by a eustatic rise in sea level.

Transport A link in the so-called morphodynamic cycle. Through transport, eroded material is moved "from A to B" and deposited when the transport energy is too low.

Transportagens (lat. *agens* = the acting, the doing), i.e. the force that brings about transport (wind, glacial ice, flowing water, marine currents).

Trough valley A glacial erosion form in the rock mass, in which detachment and detraction make a very decisive contribution.

Tundra zone Corresponds as far as possible to the subpolar zone. According to the climate classifica tion established in 1961 by W. Köppen and R. Geiger, it is the so-called "D-climate" (warmest month >10 °C coldest month < −3 °C), in the climate classification established in 1963 by C. Troll and K.-H. Paffen in 1963, it is the so-called "I2 and I3 climates" (polar and subarctic tundra climates). According to the current "effective climate classification" according to A. Siegmund and P. Frankenberg from 2014, it essentially corresponds to the "E climates" (annual mean temperature between -10 and $0\ °C$).

Tundra Term for a typical vegetation landscape in subpolar regions (periglacial areas).

Tunnel valley A form of erosion (fluvioglacial) formed under the ice ($=$ subglacial) by the erosive activity of meltwater.

Unconformity If rock formations run at an angle to each other, this is called unconformity.

Upset moraine Terminal moraine.

Urstromtal A fluvioglacial erosion form. During the cold periods, glacial stream valleys were the large receiving waters into which the numerous meltwater streams flowed as tributaries. The meltwater flowed off via the glacial valleys in the direction of the world sea. They have been preserved as valleys until today.

Valley glaciers Formed when glacial ice flows down valleys, they are narrow and elongated. They often flow downwards from higher plateaus (plateau glaciers) towards the edge. For example, the plateau glaciation of Vatnajökull on Iceland merges into various valley glaciers towards the edge.

Vegetation period The rhythmically recurring period of the year with active plant growth, which varies greatly depending on the geographical latitude. In the ever-humid tropics, for example, the growing season lasts a full 12 months, whereas in the middle latitudes (e.g. in Germany) it lasts from about mid-March to mid-October, i.e. about 7 months.

Volcanism This is a variety of different volcanic (and thus endogenous) processes, which have one thing in common, the extraction of molten rock, which rises from the Earth's mantle to the surface and is composed mainly of mantle material, possibly supplemented by crustal material melted during the ascent. Gases of varying quantities and composition are also produced.

Warm period A period of geological history spanning several thousand years that follows a cold period and is usually followed by another cold period. Currently, the Earth's history is in a warm period (the so-called Holocene). The global mean temperature is about 15 °C.

Water cycle A water cycle comprises the entire transport and storage of water in a system of varying size (e.g. global or only regional).

Weathering Collective term for all processes of so-called rock processing, i.e. processing of inorganic material. Essentially, a distinction is made between chemical processes ($=$ chemical weathering) and physical or mechanical processes ($=$ physical or mechanical weathering). Biological weathering takes place under the influence of living organisms, but either chemical (the influence of organic acids intensifies e.g. the hydrolysis) or mechanical processes (e.g. during root blasting) take place. Weathering must not be equated with decomposition, in which residual organic material is processed (mineralized). Weathering is part of the morphodynamic cycle.

Young moraine landscape Those areas in north-central Europe and in the Alpine foothills that were shaped by the inland ice of the Weichselian/Würmian cold period.

Zechstein salts During the geological epoch of the Permian (the youngest epoch of the Palaeozoic, 299–251 million years before today, roughly corresponding to the old term "Dyas") salt rocks formed in a shallow secondary sea by evaporation under arid climatic conditions, which are assigned to the younger Permian (the Zechstein, 260–251 million years before today).

References

Billwitz K et al (eds) (1992) Jungquartäre Landschaftsräume. Springer, Berlin/Heidelberg

Buchal C, Schönwiese CD (2010) Die Erde und ihre Atmosphäre im Wandel der Zeiten, 1. Aufl. Mohn Media, Gütersloh

Catt JA (1992) Angewandte Quartärgeologie. Enke, Stuttgart

Christensen S et al (1984) Geologie des Herzogtum Lauenburg. In: Degens T (Hrsg) Exkursionsführer des Nord- und Ostseeraumes. Geologisch-Paläontologisches Institut der Universität Hamburg, Hamburg, S 51–106

Claaßen C (1989) Skandinavien taucht auf. Norddeutschland ertrinkt geographie heute 10(71): 16–20

Ehlers J (1978) Die quartäre Morphogenese der Harburger Berge und ihrer Umgebung. Mitteilungen der Geographischen Gesellschaft, Bd. 68. Selbstverlag der Geographischen Gesellschaft, Hamburg

Ehlers J (1989) Eiszeiten mit Gletschern! Die Geowissenschaften 10(12):359–361

Ehlers J (2011) Das Eiszeitalter. Spektrum Akademischer Verlag, Heidelberg

Endlicher W, Gerstengarbe F-W (eds) (2007) Der Klimawandel – Einblicke, Rückblicke, Ausblicke. Potsdam-Institut für Klimafolgenforschung e. V., Potsdam

Engelhardt H (1987) Wenn Gletscher plötzlich schnell werden. Geowissenschaften in unserer Zeit 5(6):212–220

Fiedler S, Pustovoytov K (2006) Glazialer und periglazialer Formenschatz. Universität Hohenheim, Institut für Bodenkunde und Standortslehre, Stuttgart. http://www.geooek.uni-bayreuth.de/geooek/bsc/de/lehre/html/85894/glazialer_periglazialer_Formenschatz_final.pdf

Firbas F (1949) Spät- und nacheiszeitliche Waldgeschichte Mitteleuropas Bd. 1. G. Fischer, Jena

Firbas F (1952) Spät- und nacheiszeitliche Waldgeschichte Mitteleuropas Bd. 2. G. Fischer, Jena

Fischer F, Schneider H (2013) Planet 3.0: Klima – Leben – Zukunft. Schweitzerbart, Stuttgart

Flohn H (1969) Ein geophysikalisches Eiszeitmodell. Eiszeit Gegenw 20:204–231

Fraedrich W (1985) Glazialer Formenschatz in Südnorwegen. Praxis Geographie 15(5):14–21

Fraedrich W (1989a) Kaltzeiten – Markante Abschnitte der Erdgeschichte. geographie heute 10(71):4–9

Fraedrich W (1989b) Kleine Gletscherkunde. geographie heute 10(71):10–11

Fraedrich W (1995) Steht das Klima vor einer Wende? geographie heute 16(127):30–35

Fraedrich W (2015) Polarregionen – Die empfindlichen Ökosysteme geraten zunehmend in den Fukos. geographie heute 36(323):2–8

Fraedrich W (2016) Klimawandel ist keine Neuzeiterscheinung – Das Klima der Erde unterliegt einem ständigen Wandel. geographie heute 36(326):8–12

Gellert F (1958) Grundzüge der physischen Geographie von Deutschland Bd. I. VEB Deutscher Verlag der Wissenschaften, Berlin

Graß H, Klingholz R (1990) Wir Klimamacher. Fischer, Frankfurt/Main

Greiner U, Hatz G (2012) Gletscher bleiben uns länger erhalten. Kleine Zeitung vom 6. April 2012. http://www.kleinezeitung.at/nachrichten/chronik/2989657/trotz-schmelze-gletscher-halten-noch-200-jahre.story

Grube A, Wohlenberg H-J (2006) Salzstock im Buch der Erdgeschichte – Die Kalkgrube Lieth bei Elmshorn. In: Look ER, Feldmann L (eds) Faszination Geologie. Die bedeutendsten Geotope Deutschlands. Schweizerbart, Stuttgart

Hantke R (1992) Eiszeitalter – Die jüngste Erdgeschichte der Alpen und ihrer Nachbargebiete. Ecomed, Landsberglech

Hebbeln D (2014) Die polaren Meeressedimente als Archiv des Weltklimas. In: Vogt C, Lozán JL, Grassl H, Notz D, Piepenburg D (eds) Warnsignal Klima – Die Polarregionen. Wissenschaftliche Auswertungen, Hamburg, pp 254–259

Hendl M, Bramer H (eds) (1987) Lehrbuch der physischen Geographie. Deutsch, Frankfurt/Main

Intergovernmental Panel of Climate Change (Hrsg) (2015) Climate change 2014. Synthesis report. Genf

Jätzold R et al (1984) Physische Geographie und Nachbarwissenschaften. Harms Handbuch der Geographie. List, München

Kahlke H-D (1981) Das Eiszeitalter. Aulis, Köln

Kerschner H (2009) Klimawandel in Österreich. Gletscher und Klima im Alpinen Spätglazial und frühen Holozän. Alpine space – man & environment, Bd. 6. Innsbruck University Press, Innsbruck, S 5–26

Kuhle M (1988) Die reliefspezifische Eiszeittheorie. Die Geowissenschaften 6(5):142–150

Kuhle M (1991) Glazialgeomorphologie. Wissenschaftliche Buchgesellschaft, Darmstadt

Lampe R et al (2003) Nacheiszeitliche Küstenentwicklung an der Ostsee. In: Liedtke H (Hrsg) Nationalatlas der Bundesrepublik Deutschland – Relief, Boden und Wasser. Spektrum, Heidelberg, S 80. http://archiv.nationalatlas.de/wp-content/art_pdf/Band%202_80-81_archiv.pdf

Lausch E (1995) Heinrichs eisige Rätsel. Die Zeit: Nr. 12 vom 17. März 1995, S 50

Leipe T, Moros M, Tauber F (2011) Die Geschichte der Ostsee. Leibniz-Institut für Ostseeforschung, Warnemünde. http://www.io-warnemuende.de/geschichte-der-ostsee-2489.html

Leser H (1977) Feld- und Labormethoden der Geomorphologie. De Gruyter, Berlin/New York

Leser H, Panzer W (1981) Geomorphologie. Das Geographische Seminar. Westermann, Braunschweig

Lieb G-K (2004) Die Pasterze – 125 Jahre Gletschermessungen und ein neuer Führer zum Gletscherweg. http://static.uni-graz.at/fileadmin/urbi-institute/Geographie/pasterze/archiv/12 5jahre_gletschervermessung.pdf

Liedtke H (1981) Die nordischen Vereisungen in Mitteleuropa. Forschungen zur Deutschen Landeskunde, Bd. 204. Selbstverlag des Zentralausschusses für deutsche Landeskunde, Trier

Liedtke H (1986) Stand und Aufgaben der Eiszeitforschung. Geogr Rundsch 38(7–8):412–419

Liedtke H (ed) (1990) Eiszeitforschung. Wissenschaftliche Buchgesellschaft, Darmstadt

Mahlich S (2011) Klimaforschung an Eisbohrkernen – GRIP, Greenland Ice Core Project. Technische Universität Bergakademie, Freiberg. http://www.geo.tu-freiberg.de/Hauptseminar/2009/Stefanie_Mahlich.pdf

Maisch M (1989) Der Gletscherschwund in den Bündner Alpen seit dem Hochstand von 1850. Geogr Rundsch 41(9):474–482

Maisch M (1995) Gletscherschwundphasen im Zeitraum des ausgehenden Spätglazials (Egesen-Stadium) und seit dem Hochstand von 1850 sowie Prognosen zum künftigen Eisrückgang in den Alpen. In: Gletscher im ständigen Wandel. Jubiläums-Symposium der Schweizerischen Gletscherkommission Verbier (VS) "100 Jahre Gletscherkommission – 100000 Jahre Gletschergeschichte". Publikationen der Schweizerischen Akademie der Naturwissenschaften. Vdf Hochschulverlag der ETH, Zürich

Marcinek J (1984) Gletscher der Erde. Deutsch, Frankfurt/Main

May J (1988) Das große Buch der Antarktis. Maier, Ravensburg

Menke B et al (1984) Der Salzstock Lieth/Elmshorn und das Quartär von Westholstein. In: Degens T (ed) Exkursionsführer des Nord- und Ostseeraumes. Geologisch-Paläontologisches Institut der Universität Hamburg, Hamburg, pp 445–465

Meyer K-D (1990) Pleistozäne Vergletscherung in Mitteleuropa. Die Geowissenschaften 10(2): 31–36

Michler G (ed) (2010) Klimaschock – Ursachen, Auswirkungen, Prognosen. Tandem, Potsdam

Murawski H (1983) Geologisches Wörterbuch. Enke, Stuttgart

Nicolaus M, Hendricks S (2015) Das aktuelle Wissen zum Thema 'Meereis'. Bremerhaven: Alfred-Wegener-Institut (Fact Sheet). https://epic.awi.de/id/eprint/46414/1/WEB_DE_Factsheet_Meereis.pdf

Oerter H (2008) Eisbohrkerne als Klimaarchiv. Alfred-Wegener-Institut, Bremerhaven. http://epic.awi.de/21184/1/Oer2009n.pdf

Paluska A (1989) Eiszeiten ohne Gletscher? Die Geowissenschaften 10(9):258–269

Patzelt G (2008) Gletscherschwund und Vorzeitklima. Innsbruck. Bergauf 128(02):34. http://www.alpenverein.com/portal_wAssets/z_alt/portal/Home/Downloads/Bergauf_2_08/Gletscherschwund.pdf

Rahmstorf S (2004) Abrupt climate change. In: Münchener Rückversicherungsgesellschaft (Hrsg) Weather catastrophes and climate change. München. http://www.pik-potsdam.de/~stefan/Publications/Other/rahmstorf_abrupt_change_2004.pdf

Reineck H-E (1982) Das Watt – Ablagerungs- und Lebensraum. W. Kramer, Frankfurt/Main

Richter D (1980) Allgemeine Geologie. De Gruyter, Berlin/New York

Ripota P. (o. J.): Forscher haben keine Zweifel – Die nächste Eiszeit kommt bestimmt. P.M. "Perspektive Wetter und Mensch", S 78–85

Sachse W et al (1964) Steine am Strand und Kliff von Rügen. In: Gellert JF (ed) lowe und Umgebung – Nordost-Rügen. VEB Deutscher Verlag der Wissenschaften, Berlin

Schallreuter R et al (1984) Geschiebe in Südostholstein. In: Degens T (ed) Exkursionsführer des Nord- und Ostseeraumes. Geologisch-Paläontologisches Institut der Universität Hamburg, Hamburg, pp 107–148

Schulz M, Bickert T. (o. J.): Paläoklimaforschung – Rückblick in die Zukunft des Klimawandels. Fachbereich Geowissenschaften der Universität, Bremen. http://www.geo.uni-bremen.de/geomod/de/schaukasten/8_SchuBi.pdf

Schmidtke K-D (1985) Auf den Spuren der Eiszeit. Husum Druck- und Verlagsgesellschaft, Husum

Schmidtke K-D (1992) Die Entstehung Schleswig-Holsteins. Wachholtz, Neumünster

Schwarz R (1991) Klimaumschwung am Albedoübergang. Die Geowissenschaften 11(1):23–26

Smed P, Ehlers J (1994) Gesteine aus dem Norden. Bornträger, Stuttgart

Straka H (1975) Pollen- und Sporenkunde. Grundbegriffe der modernen Biologie Bd. 13. Fischer, Stuttgart

Streif H (1993) Geologische Aspekte der Klimawirkungsforschung im Küstenraum der südlichen Nordsee. In: Schnellnhuber H-J, Sterr H (eds) Klimaänderung und Küste – Einblick ins Treibhaus. Springer, Berlin/Heidelberg, pp 77–93

Stremme E et al (1984) Die Insel Sylt im Eiszeitalter. In: Degens T (ed) Exkursionsführer des Nord- und Ostseeraumes. Geologisch-Paläontologisches Institut der Universität Hamburg, Hamburg, pp 283–296

Umweltbundesamt (2006) Klimagefahr durch tauenden Permafrost. Dessau. https://www.umweltbundesamt.de/sites/default/files/medien/357/dokumente/klimagefahr_durch_tauenden_permafrost.pdf

U.S. Geological Survey Geologic Names Committee (2009) Divisions of geologic time – major chronostratigraphic and geochronologic units. Stratigraphy 6(2):90–92

Wagenbreth O, Steiner W (1985) Geologische Streifzüge. VEB Deutscher Verlag für Grundstoffindustrie, Leipzig

Washburn AL (1973) Periglacial processes and environments. Oxford University Press, London

Wehmeier E, Jakob B (2001) Die Ostfriesischen Inseln. Institut für Geographie der Universität, Stuttgart. http://www.geographie.uni-stuttgart.de/exkursionsseiten/Nwd2001/Themen_pdf/Ostfrinseln.pdf

Weise O (1983) Das Periglazial. Bornträger, Berlin/Stuttgart

Woldstedt P (1956) Die Geschichte des Flussnetzes in Norddeutschland. Eiszeit Gegenw 7(1):5–12

Zepp H (2014) Grundriss Allgemeine Geographie, 6. Aufl. Geomorphologie. Schöningh UTB, Paderborn, Stuttgart

http://www.awi.de/

http://www.biogeo.uni-bayreuth.de/biogeo/de/lehre/html/85894/glazialer_periglazialer_Formenschatz_final.pdf

http://www.biosphaere.info/biosphaere/index.php?artnr=000053

https://de.wikipedia.org/wiki/Dome_F

https://de.wikipedia.org/wiki/EPICA

https://de.wikipedia.org/wiki/Greenland_Ice_Core_Project

https://de.wikipedia.org/wiki/Greenland_Ice_Sheet_Project

https://de.wikipedia.org/wiki/Station_Dome_Concordia

https://de.wikipedia.org/wiki/Wostok-Station

http://www.diercke.de/content/jahreszeitenklimate-nach-trollpaffen-978-3-14-100700-8-228-1-0

http://www.diercke.de/content/klimate-der-erde-nach-köppengeiger-978-3-14-100700-8-229-3-0

http://www.diercke.de/content/klimate-der-erde-nach-siegmundfrankenberg-978-3-14-100700-8-226-1-0

http://earthobservatory.nasa.gov/IOTD/view.php?id=4073

http://upload.wikimedia.org/wikipedia/commons/7/7e/Milankovitch_Variations.png

http://www.geo.fu-berlin.de/v/pg-net/geomorphologie/periglazialmorphologie/Periglazialer_Formenschatz/

http://www.ipcc.ch/

http://wiki.bildungsserver.de/klimawandel/index.php/Klimaszenarien

http://www.gletscher-info.de/wissenschaft/

http://www.gletscher-info.de/wissenschaft/gletscher-landschaftsgestaltung.html

http://www.handelsblatt.com/technik/energie-umwelt/gletscherschwund-kein-schnee-am-kilimandscharo/3294318.html

http://www.krone.at/Wissen/Gletscherschwund_in_Oesterreich_wird_immer_massiver-Warmer_Sommer-Story-336370

http://www.spiegel.de/wissenschaft/natur/ausbruch-des-vulkans-toba-brachte-menschheit-dem-aussterben-nahe-a-867281.html

http://www.spiegel.de/wissenschaft/natur/klimawandel-naechste-eiszeit-kommt-in-15-000-jahren-a-303545.html

http://www.stratigraphy.org/index.php/ics-chart-timescale

Index

Printed in the United States
by Baker & Taylor Publisher Services